Physics Problems

*the text of this book is printed
on 100% recycled paper*

About the Author

Professor Clarence E. Bennett has been with the Department of Physics at the University of Maine since 1934. Previously he was a member of the faculty at Brown University and at the Massachusetts Institute of Technology. He holds the Ph.D. degree from Brown University.

A fellow of the American Association for the Advancement of Science and of the American Physical Society, and a member of the American Association of Physics Teachers, the Optical Society of America, and the American Society for Engineering Education, he has contributed articles to scientific journals and is the author of the College Outlines *Physics* and *Physics without Mathematics,* and of the standard textbook, *First-Year College Physics.*

COLLEGE OUTLINE SERIES

Physics Problems

SECOND EDITION

CLARENCE E. BENNETT

Professor Emeritus of Physics
University of Maine

BARNES & NOBLE BOOKS

A DIVISION OF HARPER & ROW, PUBLISHERS

New York, Hagerstown, San Francisco, London

First BARNES & NOBLE BOOKS edition published 1973.

LIBRARY OF CONGRESS CATALOG CARD NUMBER: 72–80370

STANDARD BOOK NUMBER: 06–460149–8

76 77 10 9 8 7 6 5 4

Table of Contents

1   Introduction   1

PART I. MECHANICS

2   The Nature of Force   12
3   Kinematics and Dynamics of Translatory Motion   32
4   Work and Energy   59
5   Rotary Motion   69
6   Periodic Motions   81

PART II. PROPERTIES OF MATTER AND SOUND

7   Statics of Elasticity and Mechanics of Fluids   88
8   Wave Motion and Sound   107

PART III. HEAT

9   Nature of Heat and Temperature   118
10   Calorimetry   125
11   Heat Transfer and Thermodynamics   132

PART IV. ELECTRICITY AND MAGNETISM

12   Electrostatics   140
13   Electric Current   159
14   Magnetism and Magnetic Effects of Currents   174
15   Electromagnetic Induction   192

PART V. LIGHT

16   The Nature of Light   210
17   Geometric Optics   220
18   Physical Optics   252

Appendix

Symbols, Concepts, and Units   256
Answers to Problems   260

Index   271

Preface

Physics is essentially a quantitative science. As science it deals with the "how" of natural phenomena, but being the science of Physics it is necessarily concerned also with the "how much." Thus problems of a numerical nature are involved in almost all aspects of a first course in Physics. The purpose of this book is to help students realize that such problems arise very naturally when relationships between physical concepts are studied, and that their solutions may be arrived at by strictly logical methods. In brief, the aim is to dissuade the beginning student from the all too common belief that a very special type of intuition is needed to cope successfully with the problem assignments which form an essential part of the properly organized first course.

To accomplish this goal, a brief summary of the topics usually covered in the first-year course is presented in a form even more condensed than that in *Physics* by the same author in the College Outline Series. Special attention is paid to the meanings of formulas, which are simply shorthand expressions for relationships between physical concepts. These formulas are illustrated by sample problems worked out in some detail, following a procedure which is based upon logical approach and careful analysis. In this manner it is hoped that the student will learn how to set up and tackle problems generally, not so much by the case method as by an analytical and systematic approach. It is the author's conviction that simply an array of solved problems will be of no avail to a student who is confronted with a problem not included in the array, whereas mastery of the correct method of approach will be helpful in solving all problems. It is indeed amazing how many elements are common to a variety of problem situations. In brief, it does seem possible to tell a student how to proceed with problem solutions instead of anticipating and showing him precisely how to solve every individual problem he might conceivably ever encounter.

In the topic summaries, definitions and defining equations in particular play important roles. This is because a clear-cut knowledge of the meaning of the concept in question is an absolutely essential first step in resolving any problem situation involving that concept. It is surprising how simple many problems become once the meaning of each concept involved is completely understood. Of course, a complete discussion of concepts intended to give the student a thorough understanding of their meanings is just what a textbook is, but this is not such a book. This book is intended as a supplement to a textbook, even a supplement to an outline of a typical textbook. The entire presentation of concepts is directed toward problem solving, yet this may not always seem obvious to the student. It is hoped, however, that once he acquires the proper viewpoint about problems as applications of the laws and principles of the sciences, the student will find this text a helpful adjunct in his problem-solving experiences. Sufficient detail is given in the definitions and relationships to provide the logical basis necessary for the approach to the problems.

Since this is a book devoted to the problem-solving aspects of the first-year Physics course, only those topics will be included which lend themselves to quantitative treatment at the first-year level. This means that no consideration will be given to such topics as electronics and nuclear Physics. In most of the elementary Physics texts these topics are treated in a more or less qualitative or descriptive manner. Questions, rather than numerical problems, are in order in these fields which, if treated quantitatively, would lead to problems somewhat beyond the scope of the first-year course. Furthermore, a background of classical Physics is so essential to the proper understanding of topics in the more modern aspects of Physics that there is ample justification for limiting the treatment here to those topics which yield problems that often seem troublesome to the first-year student.

It is not claimed that all the problems in this book are original with the author. Over the years a teacher accumulates a file of examination problems and questions whose precise origins are often difficult to trace. Many of the problems in this text owe their origin to such lists, some of which also represent a joint effort between the author and his colleagues, past and present, over a considerable period of time. For such indirect assistance from persons too numerous to mention the author is indeed grateful.

In the revised edition increased emphasis has been placed on the use of the increasingly popular mks units, but not to the exclusion of cgs and British engineering units. Also, the calculus terminology has been introduced in a few places because of its increased use in modern general physics texts. The most comprehensive change, however, is the moving electric charge approach rather than the magnetic pole approach to magnetism and the modern treatment of magnetic fields and the magnetic effects of current, all from the point of view of mks units.

1

Introduction

Physics is the study of the physical world and, as such, deals primarily with concepts. Some of these are easily recognized by the beginning student because of their everyday importance. Others are rather abstract in nature and are to be appreciated only after they have been precisely defined or derived. Even the former ones must be defined precisely to be useful in a scientific sense, in spite of the fact that precise definitions of common concepts often seem unnecessary to the beginner. Too often he is satisfied with mere qualitative expressions. It is necessary that the student understand the importance of definitions in Physics. Otherwise he will never acquire the proper "feel" for the relationships, commonly called formulas, which exist between concepts and which make possible the solving of problem situations in Physics.

The lack of this understanding often results in the view that Physics is just a collection of formulas and problems, the latter to be solved by memorizing the former and substituting values therein. Thus problem solving to many, if not most, beginning students merely requires a simple manipulation of algebra whereby one substitutes numerical values in a formula so as to compute the value of the "unknown" quantity.

Naturally such students cherish large and well-indexed collections of formulas. Indeed, they are lost if such collections do not include a formula for every conceivable problem that might be encountered. No wonder so many students become frustrated and conclude that the required first course in Physics is out of their reach. The difficulty is undoubtedly one of attitude concerning the value of definitions of physical concepts and the logical manner in which relationships between them can be expressed quantitatively as well as qualitatively.

Of course, this is not to imply that certain formulas need not be remembered. There is no substitute for memory in the learning process, but remembering is not to be confused with memorization. A student has relatively little difficulty in remembering things which he understands and with which he feels well acquainted, especially if he realizes their significance, and if, furthermore, the number of such important items can be kept to a minimum. The latter condition is particularly applicable to Physics, where the organization of the material is so complete that only a few basic relationships, or natural laws, explain a great deal. In other words, the student can easily deduce for himself many of the relationships of lesser importance from only a few basic ones if he *really* understands the meanings of the concepts involved, particularly if he has had the relationships derived for him by his instructor or seen them developed in his textbook. Thus he acquires the necessary understanding which breeds familiarity, which in turn makes it relatively difficult for him to forget the important factors involved.

Mathematical Language. As has been pointed out already, concepts are all-important in Physics. By virtue of their precise definitions, certain relationships between them logically follow. Many of these are most unambiguously expressed by the use of mathematical language, since mathematics is a form of symbolic logic. Indeed, many of the concepts are best defined by mathematical expressions called defining equations, in which single letters take the place of words and a wordy relationship is replaced by the shorthand language of symbols.

As a simple example, consider the concept of average velocity. It is defined as displacement divided by time (these concepts will be discussed later). Using the symbols $\bar{v}$ for average velocity, s for displacement, and t for time, the defining equation for average velocity becomes

$$\bar{v} = \frac{s}{t}$$

which is recognized as a mathematical expression. As a consequence it also follows that

$$s = \bar{v}t \quad \text{and} \quad t = \frac{s}{\bar{v}}$$

These expressions can be referred to quite properly as formulas for displacement and time respectively, but not as defining equations for these concepts since they are derived from the equation which defines average velocity. Thus, if the average velocity and time are specified for some physical situation, it is hardly a matter of memorizing a formula to determine the displacement, since there are many formulas expressing displacement in terms of concepts other than velocity and time. If one understands exactly what is meant by the concept of average velocity, i.e., if one knows that $\bar{v} = s/t$ by definition, it readily follows that $s = \bar{v}t$, without any necessity for memorizing the relationship.

To be sure, this has been a very simple illustration, but the principle applies in varying degrees to practically all formula situations. It is not that some kind of formula may not be necessary to solve a given problem, but that the basic relationship between the concepts is more important. From a basic relationship which is often simply a defining equation, the student can express the unknown in terms of the known quantities by relatively simple manipulations. Obviously, the more complex the situation is, the more necessary it is to understand the basic nature of each concept entering into a given relationship in order to manipulate it successfully for the desired answer.

It might even be argued that all one really has to know are the concepts, and the relationships will take care of themselves. This, however, is unrealistic for the beginning student. Hence a limited number of the more important formulas are well worth remembering, but only to the extent that they convey meaning and are not memorized in parrotlike fashion.

Thus is it seen that mathematics becomes a shorthand language in Physics. It replaces cumbersome words for the person who understands what the symbolism stands for, but is obviously worse than meaningless, if not outright confusing, to the person who attempts merely to memorize the shorthand expressions for their own sake. It is truly a means to an end and must under no circumstances become the end in itself, particularly at the elementary level.

One difficulty which is cleared up for the student, once he appreciates the value of definitions or defining equations, is the looseness with which common technical terms are used by the layman. Although the beginner is himself often satisfied with loose usage of words, he soon discovers that such looseness complicates enormously the task of keeping relationships straight. Indeed, he may be apt to conclude that the situation is hopeless and seems to require the memorization of an enormous number of trivial details. If he stays with the task long enough, however, he comes to appreciate the fact that careful usage of properly defined technical terms really simplifies the situation rather than complicates it, as he at first thought. Unfortunately, most Physics courses must proceed so rapidly that many students never achieve this result. It is hoped that this text will assist such students to appreciate the proper viewpoint and thereby be of value as a supplement to the regular textbook used in Physics courses.

The Importance of Problems in Physics. In the preceding section the point was made that when concepts are precisely defined, relationships between concepts necessarily arise, and that mathematical methods can be employed to express these relationships. It should now be clear to the student that by manipulating these mathematical expressions certain deductions concerning the concepts can often be made which might not at first be apparent. Thus relationships can be derived from other relationships until ultimately quite a substantial body of knowledge can often be accumulated around a set of concepts. Obviously these relationships will be recognized as formulas which can be utilized to answer questions that seem to arise about natural phenomena. Thus so-called problem situations develop all through the study of Physics. Indeed, these problems constitute the life blood of the science so that the science of Physics may even appear to be nothing but a collection of problems. Therefore, the student of Physics must not only be prepared to solve problems, particularly problems formulated in mathematical language, but must have the desire to do so if he is to progress beyond the definition stage.

It has been said that unless one can cope with the quantitative aspects of physical relationships, i.e., unless one can answer the "how much" as well as the "how," one does not know very much about the subject of Physics. Consequently problems in Physics are numerical, and certainly this suggests mathematics. Yet it seems worth repeating that mathematical manipulation without any regard for what the mathematical symbols stand for is not to be thought of as Physics. In other words, merely poring through a set of formulas for just the one which expresses the sought-for quantity and then simply substituting numerical values and solving for the unknown quantity by arithmetic is hardly the proper goal of the serious Physics student.

The real problem is understanding the physical situation, recognition of concepts involved, recollection of precise definitions of those concepts, and manipulation of symbolic language used to express relationships between them so as to result in an intelligent answer to specific questions asked about them. This is the challenge to the student, and it is one which is welcomed by the serious student because he realizes that the more he is questioned about the concepts of the science the more he learns about them.

Thus problems that are customarily found in first-year Physics courses are intended to be aids in the learning process in spite of the fact that many students believe them to be included for the sole purpose of tripping them up. Indeed they are opportunities for the student to test his grasp of the subject matter, and obviously are not to be tackled before the subject matter is studied. Since the problems represent a tangible aspect to the lesson assignment, there seems to be a tendency for some students to "get the problems over with" before even reading the lesson itself. Of course this is precisely the wrong approach. Until the student can truthfully say that he thinks he understands the subject matter at hand and is ready to be tested on it, he is not ready to try the problems. Thus when he follows the correct method just outlined, the problems often cause him little if any difficulty.

Basic Concepts vs. Derived Concepts. In considering the concepts which make up the science of Physics, it soon becomes evident that certain concepts are more basic, or fundamental, than others. As was pointed out in the preceding sections, each new concept to be studied must either be defined or be derived from earlier ones if the study is to have meaning. This suggests that there must be a beginning somewhere, i.e., that somewhere early in the process a certain few of these concepts must be accepted as incapable of definition in terms of anything simpler. Indeed, it is found that from only *three* such concepts all of the other concepts which are encountered in that large branch of Physics known as Mechanics can be so defined or derived. These are *length, mass,* and *time.* They must be accepted by the student as simply words or terms which have common meanings to all people without further elaboration. In other words, length is simply length, mass is simply mass, and time is simply time. Further explanation of these concepts is not possible. A little reflection shows that it is really not the concept of length which bothers a student. It is rather the question of how long an object is. The same can be said for mass and time. Once this is agreed to, the stage is then set for a consideration of all the other concepts which follow from these three. Thus average velocity, $\bar{v}$, previously mentioned and to be considered in more detail later, is definable as

$$\bar{v} = \frac{s}{t}$$

where s is essentially a length and t stands for time. Similarly such a concept as momentum (more about this concept later) is definable as the product of mass and velocity, or mv. Also kinetic energy will be defined later as

$$KE = \tfrac{1}{2}mv^2$$

These are examples of so-called derived concepts, of which there are a great many in Physics.

Standards and Units. Referring back to the matter of determining the length of an object, it is clear that a unit of length must be established in which to express the length of anything in a quantitative manner. Also, because only the three concepts of length, mass, and time are the basic ones, it follows that only three basic units of measure are necessary to express the amount of length, the amount of mass, and the amount of time in any or all mechanical situations.

The establishment of such units is a matter of common agreement and enforcement. At an early date, the distance between two scratches on a certain bar of platinum iridium was recognized as a standard unit of length known as the *International Standard Meter* in the metric system of units. This distance was redefined in 1960 in terms of wave lengths of light to give it a more natural basis, but the specified platinum-iridium bar is still the practical standard. Another such standard is the *British Yard* in the English system. The meter (m) is subdivided into 100 equal parts or centimeters (cm), each of which is the equivalent of 10 millimeters (mm). The yard comprises 3 ft, each of which is equivalent to 12 in. The conversion factor between the two systems is given by the relation that

$$1 \text{ yard} = \tfrac{36}{39}\text{m}$$

or
$$1 \text{ in.} = 2.54 \text{ cm}$$

The standard of mass in the metric system is a certain block of platinum referred to as the *kilogram* (kg), the one-thousandth part of which is the gram (g). In the English system the unit of mass is the *pound mass* containing 16 ounces.

$$1 \text{ lb} = 454 \text{ g} = .454 \text{ kg}$$

The standard unit of time in the metric, as well as the English, system is the mean solar day, i.e., the time required for the earth to rotate once about its axis, averaged over the entire year and measured with respect to the sun.

$$1 \text{ day} = 24 \text{ hours} = 24 \times 60 \text{ minutes} = 24 \times 60 \times 60 \text{ seconds}$$
$$\therefore \quad 1 \text{ day} = 86,400 \text{ seconds}$$

The student is assumed to have the arithmetic ability to convert from meters to centimeters or millimeters, or from pounds to ounces or even tons, and from centimeters to inches to feet, and from grams to pounds, and from seconds to minutes or hours, almost without thinking. It is needless to illustrate this procedure by problems because of the relative simplicity of the operations. Yet a certain number of illustrative problems will be generally used in this book to help familiarize the student with the concepts as they are introduced.

Comments Regarding the Numerical Aspects of Problems. In the problem work which follows in this text it is suggested that all calculations be made with the slide rule. By this means not only will time be saved, but also an appreciation of significant figures will be cultivated. No numerical answer to a problem in Physics is more accurate, or precise, than the values given as data. Thus if a length is specified as 10.1 cm, for example, no calculation involving this length should be carried beyond three significant figures. This is because no digit beyond the third in such a calculation is significant even if the arithmetic is flawless.

The carrying out of answers beyond three figures serves only to create a *false* impression of accuracy. In fact if the problem involves other data which are expressed to fewer than three significant figures—suppose time enters in and is given as 3.4 sec— the answer is correct to only the significance of the least precisely given data. In this case, it would be senseless to express an answer for average velocity as

$$\bar{v} = \frac{s}{t} = \frac{10.1}{3.4} = 2.970 \text{ cm/sec}$$

It should be 3.0 cm/sec. Obviously, slide rule accuracy is adequate for such calculations because experimental data (usually obtained by measurements) are ordinarily not expressed to greater accuracy than three figures.

If one is to be meticulous in the matter of significant figures, one follows the rule of carrying the calculations one figure beyond the one which is certain. Consequently the last figure written for any quantity is always treated as uncertain. Thus if data are expressed to three figures, the reader always interprets this to mean that only two figures are really significant. Therefore, in first-year Physics problems, where an understanding of the laws and principles of the subject are more important than the arithmetic, there is ample justification in limiting answers to two or three digits, as the case dictates.

In this book, answers to all problems are provided in the Appendix. Students are warned, however, that apparent discrepancies between their own answers and those in the book may occur in the third significant figure.

It should now be obvious to the student that, insofar as Physics is concerned, as distinguished from Engineering applications, a numerical answer to a problem is less important than an algebraic answer, which is a general answer. The numerical answer, which may be of considerable practical importance, is a specific matter which is usually accomplished by a single slide rule operation involving numbers only after the general (algebraic) answer is obtained by a logical manipulation of the concepts involved in the problem situation. Hence all problems should be solved algebraically before the numbers are introduced as a final single step in the solution.

Part I.

Mechanics

2

The Nature of Force

Although it is commonly recognized that what is known as Physics comprises such topics as mechanics, properties of matter, wave motion, sound, heat, electricity, magnetism, electronics, and light, it is not generally realized by the beginner that all these topics are closely interrelated. Indeed, Physics is probably the most highly organized of all the sciences, so that it becomes very important to start at the proper place. It turns out that *mechanics,* which is the study of forces and motion, is basic to all the others and that the order of topics just given is a proper order in which to study them.

The two concepts, *force* and *motion,* are so completely interrelated that neither can be defined independently of the other. It is customary first to consider some aspects of the concept of force, which is a more or less familiar concept, and then later to develop the ideas associated with motion.

Referring to common experience, and not considering all the implications suggested by the term motion, force is nevertheless recognized as *a push or a pull tending to change the motion of a body.* It is measured in pounds (English units) or newtons (metric units) or dynes (also metric units). Force is a concept that will be found later to be related to mass; hence it is not to be thought of as a basic concept. On the other hand, until this relationship is described in Chapter 3, the concept of force will be treated as a fourth basic concept in order that advantage may be taken of the familiarity which the student has with it. The relationship between the various force units will be discussed later.

Vector Aspect of Force. Force, along with many other concepts in Physics, has, in addition to magnitude, a directional characteristic. Quantities which require a specification of direction as well as magnitude are called *vector quantities*, in contrast to *scalar quantities*, which require a specification of magnitude only. Vector quantities can be represented by directed arrows. The length of the arrow represents, to scale, the magnitude or the amount of the quantity; the direction of the arrow, measured as an angle with respect to a given direction, represents the direction of the quantity. A force then, is represented as in Fig. 1, where the length (here 6 units) may represent 6 lb (or 6 of any other unit) of force, and the direction is indicated by the angle θ, measured with the horizontal. (Greek letters are usually used to represent angles.)

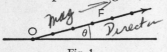

Fig. 1

It is clear that the study of forces becomes a geometric study, and the laws of geometry and trigonometry also apply, since arithmetic and algebra are insufficient to deal with directed quantities.

Weight as a Force. *Weight* is the pull of the earth on a body, sometimes called the pull of gravity, and is representable by a vector arrow pointing downward. There is a point called the *center of gravity* at which the weight of a body can be assumed to act. Since weight is a force, it must not be confused with the concept of mass; yet this error is often made by the beginning student because in lay language the terms weight and mass are frequently used interchangeably. Here is a good example of how loose usage of technical terms by the layman makes the study of Physics needlessly difficult. The relationship between weight and mass will be explained later.

Addition of Forces. Within the precision of measurement by rulers and protractors, forces can be added graphically simply by laying off vector arrows to represent the forces in question. For example, force F_2 can be added to force F_1 by first laying off F_1 and then from the head end of F_1 laying off F_2 as in Fig. 2.

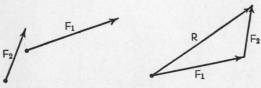

Fig. 2

A straight line drawn from the tail end of F_1 to the head end of F_2 and pointing as indicated, represents the vector sum, or the resultant, R, of F_1 and F_2. It should be noted that R is also the diagonal of a parallelogram that could be drawn with a pair of sides F_1 and a pair of sides F_2 as in Fig. 3.

Obviously, from considerations of geometry and trigonometry it is possible to calculate the length and the direction of the R vector in terms of the lengths and directions of the F_1 and F_2 vectors. Consequently the forgoing, which is a graphical method only of adding two vector quantities, can be supplemented by a mathematical method.

Resultant of Vectors by Method of Components. Consider a force represented by the vector F as in Fig. 4. Through the base O of the vector draw conventional mathe- matical x- and y-coordinates as shown. Consider the angle between F and the positive X-axis to be θ. Drop a perpendic- ular from the head end of F to the X-axis. Designate by F_x the projection of F on the X-axis. Similarly designate by F_y the pro-

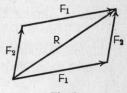

Fig. 3

jection of F on the Y-axis. In the triangle OAB note that AB is the same length as F_y. Thus F can be treated as the resultant of F_x and F_y. Furthermore, since the X- and Y-axes make 90° with each other, F_x and F_y are mutually perpendicular. Since the triangle

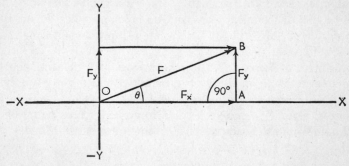

Fig. 4

OAB is a right triangle, it follows that $F^2 = F_x{}^2 + F_y{}^2$ (Pythag- orean theorem).

This means that if the x- and y-components of a resultant vector can be found, then the resultant vector itself can be computed from the Pythagorean theorem. Thus a method is suggested for calculating the resultant of two vectors instead of determining the resultant by graphical methods. The method consists of several steps. In the case of forces F_1 and F_2, first break each force into its x- and its y-component.

$$F_{1x} = F_1 \cos \theta_1 \quad F_{1y} = F_1 \sin \theta_1$$
$$F_{2x} = F_2 \cos \theta_2 \quad F_{2y} = F_2 \sin \theta_2$$

$$R_x = F_{1x} + F_{2x} = F_1 \cos \theta_1 + F_2 \cos \theta_2$$
$$R_y = F_{1y} + F_{2y} = F_1 \sin \theta_1 + F_2 \sin \theta_2$$

Adding the x-components of the forces gives the x-component R_x of the resultant, and similarly adding the y-components of the forces gives the y-component R_y of the resultant. Moreover, $R_y/R_x = \tan \phi$ where ϕ is the angle which the resultant makes with the horizontal. See Fig. 5.

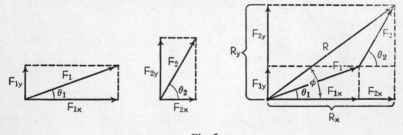

Fig. 5

$$R^2 = R_x{}^2 + R_y{}^2$$
or $$R = \sqrt{R_x{}^2 + R_y{}^2}$$

Comment. The method of components has been treated in detail because it is of basic importance to the later problem work which is the theme of this book.

Interpretation of Resultant. It is important to note that the resultant of two forces (or vectors generally) is a concept that is understandable only through the use of a vector diagram where each line has significance. A graphical construct makes it possible to evaluate the resultant of two forces, but a mathematical solution, such as is obtained by the component method, is considered to give more precise values for the magnitude and the direction of the resultant. Thus the mathematics supplements rather than replaces the graphical method. The student should always make a graphical sketch to get a general appreciation of the problem and a rough idea of the answer for the purpose of checking the answer which is obtained by the mathematical method of components.

Problem

A sled is being pulled by a cord making an angle of 30° with the ground. The tension in the cord is 50.0 lb. What is the component of the tension parallel to the ground? What is the component of the tension perpendicular to the ground?

Solution

First: Draw a diagram (Fig. 6).

Second: Draw a vector to represent the tension T, and indicate the horizontal and vertical components of the tension by drawing X- and Y-axes through the base of the vector (Fig. 7) and dropping perpendiculars to them.

$$T_x = T \cos 30°$$
$$T_y = T \sin 30°$$

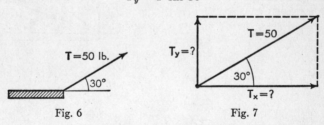

Fig. 6   Fig. 7

Third: Substitute numerical values and solve for numerical answers.

$$T_x = T \cos 30° = 50(.866) = 43.3 \text{ lb (parallel)}$$
$$T_y = T \sin 30° = 50(.500) = 25.0 \text{ lb (perpendicular)} \quad Ans.$$

Force of Friction. If a body slides on another body, a force called the force of friction is developed where the surfaces of the two bodies are in contact. The direction of this force is tangent to the surface of contact and opposite to the motion. It is also found by experience that the magnitude of the frictional force is dependent upon the perpendicular force pressing the two bodies together as well as upon the nature of the surfaces in question.

Law of Friction. If the perpendicular component of the force pressing two surfaces together, designated as N (normal force), is divided into the tangential component of the same force (the friction force F itself), the ratio is found experimentally to be constant for a given pair of surfaces.

$$\frac{F}{N} = n \quad \text{or} \quad F = nN$$

where n is called the *coefficient of friction* between the two surfaces. This is known as the law of friction.

Problem
A crate is dragged along the ground by a rope parallel to the ground. If the tension in the rope is 20 lb and the coefficient of friction between crate and ground is .40, what is the weight of the crate?

Solution
First: Draw a diagram (Fig. 8).
Second: Recall the law of friction $F = nN$.
Third: Realize that in this situation $W = N$ and $T = F$.
Fourth: Express in terms of the unknown, i.e., $W = N = F/n$.
Fifth: Substitute numerical values and solve for numerical answer.

$$W = \frac{20}{.40} = 50 \text{ lb} \quad Ans.$$

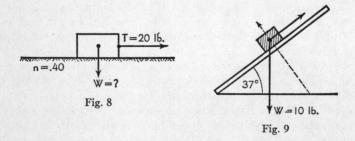

Fig. 8

Fig. 9

Problem
A box weighing 10 lb is pulled slowly and at constant speed up an inclined plank which makes an angle of 37° with the horizontal, by a

cord parallel to the incline. The resultant force along the incline is 6 lb. What is the coefficient of friction between box and plank?

Solution

First: Draw a diagram indicating the forces acting (Fig. 9).

Second: Recall that coefficient of friction is given by the defining relation $n = F/N$, where F is the resultant force parallel to plane and N is force perpendicular to plane.

Third: Determine component of W perpendicular to plane. Note that angle between W and the perpendicular to plane is same as angle between plane and horizontal.

$$\therefore\ N = W \cos 37°$$
$$= 10(.800) = 8\ \text{lb}$$

Fourth: Substitute numerical values and solve for n.

$$n = \frac{F}{N} = \frac{6}{8} = .75 \quad Ans.$$

Problems: Resultant of Vectors and Forces

1. Two forces act through the same point. The first has a magnitude of 20 lb and acts 30° above the horizontal and toward the right. The other has a magnitude of 30 lb and acts 37° below the horizontal and toward the right. (a) Diagram these forces on an xy coordinate plane. (b) Determine the sum of the x-components of these two forces. (c) Determine the sum of the y-components of these two forces. (d) Determine the magnitude of the resultant of these two forces.

2. Two forces, one of 90 lb and one of 120 lb, are acting on a body, the 120 lb force acting horizontally toward the right and the 90 lb force acting vertically upward. What is the magnitude of their resultant?

3. A force of 10 lb acts toward the east. Another force acts at the same point so that the resultant of the two forces is zero. What is the magnitude of the second force?

4. A 20 lb block is being drawn at constant speed along a rough horizontal surface by a force of 4 lb. (a) What is the magnitude of the "normal force"? (b) What is the magnitude of the force of friction? (c) What is the value of the coefficient of sliding friction between these two surfaces?

5. It requires a horizontal force of 150 lb to just start a 100 lb box

moving along a floor. What is the coefficient of static friction?

6. Compute the component of a 10 lb force in a direction which makes an angle of 60° with itself. What is the component perpendicular to itself?

7. A child's wagon is pulled by the handle which makes an angle of 30° with the horizontal. What is the component of the pull parallel with the ground if the pull along the handle is 20 lb? What is the lift?

8. Three forces A, B, and C act on a given point and make angles of 120° with one another. If the magnitude of each force is 10 lb, what is the magnitude of the resultant force?

9. A body is acted on by two forces, A and B, which make an angle of 90° with each other. The magnitude of A is 20 lb, and that of B is 30 lb. What is the magnitude and the direction of the resultant?

10. A 5 lb block is pushed along a horizontal surface by a stick which makes an angle of 60° with the horizontal. Assuming the force of the stick acts along the direction of the stick and amounts to 2 lb, what is the resultant vertical force acting on the block?

Concurrent Forces in Equilibrium. If several forces act simultaneously on a body, they may or may not act on the same point of the body. If they do act on the same point, or if their lines of action intersect at some common point, the forces are said to be *concurrent*.

The resultant force action on a body determines the motion of the body, of which there are two kinds, translation and rotation. *Translatory motion* is characterized by the fact that all straight lines remain parallel during the motion. *Rotary motion* of a body is such that every point of the body revolves in a circular path about some point, not necessarily in the body, which serves as the axis of rotation. At this point the student is reminded that he cannot hope to know all about force until he knows all about motion. Yet, without going further at this time into the terminology and concepts of motion, much can be learned about force by realizing that the net force action on a body which is at rest, or is experiencing no change in motion, must be zero. Such a situation is known as *equilibrium*.

First Condition of Equilibrium. For a body to be in equilibrium the first condition that must be met is that the resultant force acting on the body (same as the sum of the forces and represented as ΣF) must be zero, i.e., $\Sigma F = 0$. This also means that the sum of the x-components must be zero, and the sum of the y-components must be zero for a body at rest or moving at a constant rate (no change in motion).

$$\begin{cases} \Sigma F_x = 0 \\ \Sigma F_y = 0 \end{cases}$$

If the forces are concurrent, this is the only condition that must be met. Situations involving sets of forces acting on bodies at rest (statics) are very common, and it becomes possible by the use of the above relationship to calculate very easily the values of certain forces in a set when information about only some of them may be given.

Problem Procedures

The solution to any problem in statics involving concurrent forces is found simply by applying the information concerning forces given in the preceding paragraphs. Certain straightforward logical steps should be followed.

1. The problem should be read carefully and studied to the extent that the problem situation can be visualized and then clearly expressed by a simple diagram, which should then be drawn. All the known factors should be indicated on the diagram, and all the quantities to be evaluated (unknowns) should be recognized and indicated as unknowns on or near the diagram. Indeed, this first step is not limited to problems in statics but is common to all problem solving in Physics. It is really the key to success and should be practiced by the student until he finds himself doing it automatically.

2. The body, or the portion of the body which is in equilibrium, should be "isolated" for a detailed consideration of the forces acting *on* it. The reason for this is that the motion of a body is determined by the force action on it, and if a body or a portion of a body is in equilibrium, and if the forces are concurrent, then the resultant force is zero. This information helps to evaluate one or more of the forces which may not be specified in the problem. The "isolation" is carried out by actually drawing a line on the diagram (preferably with colored pencil) completely around the body or portions of the body in question. If

the forces in the problem are concurrent, the point of intersection of their lines of action is the logical portion of the body to isolate.

3. Indicate by arrows all the forces that act *on* the isolated body, including the unspecified ones (unknowns) as well as the known ones, being sure not to include forces exerted *by* the body on something else, since it is the resultant force action *on* a body which determines its motion. These arrows will all cross the colored line which "isolates" the body. Incidentally, it will be noted that for every force acting *on* the body there will be an equal but opposite force exerted *by* the body. Disregard the latter.

4. Treat these arrows as a vector diagram. Often it is convenient to redraw the arrows in a supplementary diagram where they can be sketched roughly to scale in those cases where the magnitudes and directions are known. This will make it possible to visualize roughly the components of the forces since these enter into the next step, involving the process of setting up the resultant force, whose actual value is known to be zero because of equilibrium, but whose component parts are not necessarily separately equal to zero.

5. Evaluate and tabulate the x-components of all the forces acting on the isolated body. It is recommended that the beginner construct a table somewhat as follows:

FORCE	X-COMPONENT	Y-COMPONENT
F_1	$F_1 \cos \theta_1$	$F_1 \sin \theta_1$
F_2	$F_2 \cos \theta_2$	$F_2 \sin \theta_2$
F_3	$F_3 \cos \theta_3$	$F_3 \sin \theta_3$

6. Sum up the x-components listed in the table to express the x-component of the resultant force, whose value is known to be zero.

$$\Sigma F_x = F_1 \cos \theta_1 + F_2 \cos \theta_2 + F_3 \cos \theta_3 + \ldots = 0$$

Following this the y-components are similarly summed up.

$$\Sigma F_y = F_1 \sin \theta_1 + F_2 \sin \theta_2 + F_3 \sin \theta_3 + \ldots = 0$$

It will be noted that this step gives two simultaneously true equations in terms of quantities some of which are known and some of which are to be calculated. If no more than two of these quantities are unknown the two equations can be solved for their values by algebra.

7. Solve the two equations for no more than two unknowns. In general it is well to keep the numerical work to a minimum by solving algebraically as far as possible before substituting numerical quantities. A minimum number of slide rule operations suffices to give the desired answers if this is done. If it should appear that insufficient data are given, i.e., if there appear to be more than two unknowns, look for information that may be given indirectly rather than directly. If, for example, friction forces are involved, recall the law of friction $F = nN$. Values of n and N may be given, from which F may be evaluated and then treated as a known quantity.

Positive and negative signs used with the forgoing vector concepts specify the directions of the vector quantities, i.e., up is positive, down is negative, to the right is positive, and to the left is negative.

Problem

A 5.00 lb weight suspended from a 3.00 ft cord is pulled to one side by a horizontal force F so that the cord makes an angle of 30° with the vertical. What is the value of the force F, and what is the tension in the cord?

Solution

Step 1: Diagram (Fig. 10).
Step 2: Isolate the body.
Step 3: Indicate forces *on* the body.
        These are T, F, and W.
Step 4: Consider vector diagram (Fig. 11)
Step 5: Construct table of force
        components.

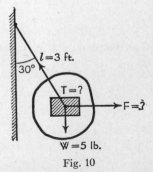

$l = 3$ ft.

$T = ?$

$F = ?$

$W = 5$ lb.

Fig. 10

FORCE	X-COMP.	Y-COMP.
T	$-T \cos 60°$	$+T \sin 60°$
F	$+F$	0
W	0	-5

Step 6: $\Sigma F_x = 0;\ \Sigma F_y = 0$.

$$\left\{\begin{array}{l} -T\cos 60° + F = 0 \\ T\sin 60° - 5 = 0 \end{array}\right\}$$

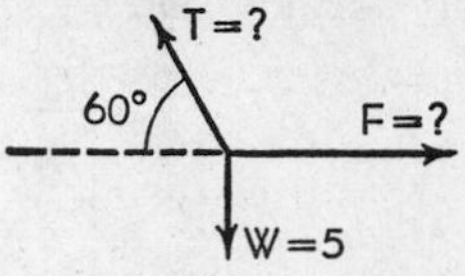

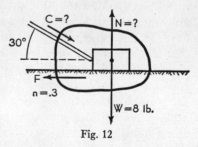

Fig. 11

Step 7: Solve the equations for the unknown quantities.

$$\left\{\begin{array}{l} -T(.5)\quad + F = 0 \\ T(.866) - 5 = 0 \end{array}\right\}$$

$$\left.\begin{array}{l} T = 5/.866 = 5.77\ \text{lb} \\ F = (5.77)\ (.5) = 2.89\ \text{lb} \end{array}\right\}\ \textit{Ans.}$$

Problem

A box is pushed along a horizontal floor by a stick making an angle of 30° with the horizontal. The coefficient of friction between box and floor is .3. If the box weighs 8.00 lb, what is the compression in the stick and what is the normal (perpendicular) force of the floor on the box?

Solution

Diagram (Fig. 12).
Isolate the body.

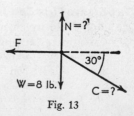

Fig. 12

Construct a vector diagram (Fig. 13).

Fig. 13

Tabulate x- and y-components of the forces acting on the body.

FORCE	X-COMP.	Y-COMP.
N	0	$+N$
C	$C \cos 30°$	$-C \sin 30°$
W	0	-8
F	$-F$	0

Utilize condition of equilibrium.

$$\begin{aligned} \Sigma F_x = 0 \\ \Sigma F_y = 0 \end{aligned} \left\{ \begin{aligned} C \cos 30° - F = 0 \\ N - C \sin 30° - 8 = 0 \end{aligned} \right\}$$

Solve for the unknowns, utilizing the law of friction $F = nN$.

$$\left\{ \begin{aligned} .866C - .3N = 0 \\ N - .5C - 8 = 0 \end{aligned} \right\} \begin{aligned} N = 8 + .5C \\ .866C - .3(8) - .150C = 0 \\ .716C = 2.4 \end{aligned}$$

$$C = \frac{2.4}{.716} = 3.35 \text{ lb}; \ N = 8 + .5(3.35) = 9.68 \text{ lb} \quad Ans.$$

Note! Since the problems illustrated here and those included in this section and the next section involve only the composition and resolution of forces, it does not matter what system of units is used. It is assumed at this stage of the study that students using this text are most familiar with the English unit of force, the pound. In each problem, however, an appropriate metric unit could be substituted without altering the problem in any way. Later on, in problems involving relationships between force and other concepts, units must be used with great care. Both metric and English units will then be used, including the increasingly popular mks units which will be introduced in Chapter 3.

Problems: First Condition of Equilibrium

1. A 1000 lb weight on the end of a cable is held out so that the cable makes an angle of 30° with the vertical. It is held in this position by a horizontal rope tied to the weight. (a) Diagram the forces (acting on the weight) involved in this problem on an xy coordinate plane. (b) Make a table of the x-components and the y-components of all the forces. (c) Determine the magnitude

of the force exerted by the horizontal rope. (d) Determine the magnitude of the force exerted by the supporting cable.

2. A 50.0 lb block is being drawn at constant speed along a rough floor by a force of 20.0 lb exerted on a rope which makes an angle of 30° above the horizontal. (a) What is the magnitude of the normal force? (b) What is the magnitude of the force of friction? (c) What is the value of the coefficient of friction between these two surfaces?

3. A 20.0 lb block is being pulled up a 30° inclined plane at constant speed by a rope which is parallel to the plane. A force of friction of 2.5 lb acts parallel to the plane, opposing the motion. (a) Draw a diagram showing all the forces which act on the block. (b) Make a table of components, showing all the x-components and all the y-components of the forces acting on the block. (c) Determine the magnitude of the normal force of the plane on the block. (d) Determine the magnitude of the pull of the rope on the block. (e) What is the magnitude of the resultant force on the block?

4. A small iron ball weighing 2.00 lb is suspended by a cord and is held at an angle of 30° from the vertical by a horizontal force F. (a) Indicate by arrows and labels on the diagram all the forces acting on the ball. (b) Does the pull of the cord on the ball act *up* along or *down* along the cord? (c) Tabulate the x- and y-components of all the forces acting on the ball. (d) Calculate the magnitude of the force F.

5. In order to move a 150 lb crate up a 30° incline and onto a platform a man pulls with a force of 100 lb. This force is just sufficient to slide the crate at constant speed. (a) Indicate by arrows and labels on the diagram all the forces acting *on* the crate. (b) Does the friction force act up along or down along the incline as the crate moves upward? (c) Tabulate the components of all the forces acting on the crate and perpendicular to the incline (i.e., take x-components along and y-components perpendicular to incline). (d) Calculate the magnitude of the friction force. (e) Calculate the coefficient of friction between crate and incline.

6. A person who weighs 180 lb sits in a hammock whose ropes make angles of 30° and 45° respectively with the vertical. What is the tension in each rope?

7. A 40 lb weight suspended from a cord is pulled horizontally by a force F to such a position that the supporting cord makes an angle of 60° with the vertical. If it is held stationary in this position, what is the magnitude of the force F?

8. A rope attached to the front end of a stalled automobile is fastened to a tree 20 ft away. A person grasps the rope at the midpoint

and pushes it in a direction perpendicular to itself. When the slack is taken up the midpoint of the rope is found to have been displaced 1 ft. If now the person continues to push with a force of 50 lb, what force is exerted on the car by the rope? In other words, what is the tension in the rope?

9. A 60 lb weight hangs from the end of a horizontal strut protruding 8 ft from a vertical wall. A tie rod connects the outer end of the strut to a point in the wall 4 ft above the strut. Find the compression in the strut and the tension in the tie rod.

10. A picture is supported on a wall by a cord connected between two eyelets, one on either side of the picture, and hanging over a peg. At the peg, the cord on either side makes an angle of 30° with the horizontal. What is the tension in the cord if the picture weighs 4 lb?

Nonconcurrent Forces and Torques. In the event that the forces acting on a body are nonconcurrent, the sum of the forces may be zero and yet the body may not be in equilibrium. This is because of possible rotational tendencies even though translational tendencies are balanced. If there is an axis of rotation associated with a body (the axle of a wheel, for example), a force perpendicular to it and so directed that its line of action does not pass through it (see Fig. 14) will tend to produce clockwise or

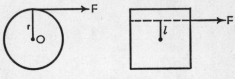

Fig. 14

counterclockwise rotation about the axis. It is found experimentally that such a tendency is not only proportional to the magnitude of the force, but is also proportional to the perpendicular distance from the axis to the line of action of the force (r in the diagram to the left where the force is tangential, but l in the diagram to the right). This perpendicular from the axis to the line of action of a force is called the *lever arm* of the force. The product of the force and its lever arm is called *torque*, or *moment of force*. The lever arm is sometimes called *moment arm*. Whereas forces tend to produce changes in translatory motion, torques tend to produce changes in rotational or rotary motion.

Second Condition of Equilibrium. If a body is in equilibrium, not only must the forces acting on it add up to zero but also all torques exerted by these forces with respect to any axis whatever must also add up to zero. This is known as the second condition of equilibrium:

$$\Sigma L = 0$$

where L stands for torque.

Problem Procedures

If a problem involves nonconcurrent forces in equilibrium, it will be necessary, in addition to the forgoing steps listed for concurrent force problems, to consider all the torques. This will first necessitate the determination of the lever arm associated with each force, with respect to some arbitrarily chosen axis. Each lever arm must then be multiplied by its respective force to evaluate each torque. Torques will be *clockwise* (negative) or *counterclockwise* (positive). When added together their algebraic signs must be taken into consideration. It will be well to add a third column to the force component table to contain the torque values. A summation of these values will obviously give a third simultaneous equation, which consequently increases to three, the number of unknowns capable of evaluation by the methods of algebra.

Incidentally, by choosing the possible axis of rotation at different places, an indefinite number of equations can be developed by summing torques, whence the number of possible unknowns capable of being determined becomes indefinite. By now the student should appreciate the value of the above method of analyzing problems, since it provides a very considerable amount of power in figuring out an indefinite number of otherwise unknown quantities in various problem situations.

Problem

A uniform plank 10 ft long (center of gravity 5 ft from either end) rests against a smooth (frictionless) wall making an angle of 53° with the floor. The plank weighs 40 lb. What horizontal force does the wall exert on the upper end of the plank? What are the vertical and horizontal components of the force exerted by the floor upon the lower end of the plank?

Solution

Diagram (Fig. 15).
Isolate the plank.
Indicate all forces acting on the plank.
Tabulate x- and y-components of these forces, and tabulate torques about base O.

Fig. 15

FORCE	X-COMP.	Y-COMP.	FORCE $\times$ LEVER ARM ABOUT O
P	$-P$	0	$+P \times l \sin 53°$
W	0	-40	$-40 \times l/2 \cos 53°$
H	$+H$	0	$H \times 0 = 0$
V	0	$+V$	$V \times 0 = 0$

Set $\begin{aligned} \Sigma F_x &= 0 \\ \Sigma F_y &= 0 \end{aligned}$ $\left\{ \begin{aligned} -P + H &= 0 \\ -40 + V &= 0 \end{aligned} \right\}$ $\therefore$ $\begin{aligned} V &= 40 \text{ lb} \quad Ans. \\ H &= P = ? \end{aligned}$

Set $\Sigma L = 0$ $\quad P(10)(.8) - 40(5)(.6) = 0$ $\quad \therefore$ $\quad P = \dfrac{40(5)(.6)}{10(.8)}$

$$= 15 \text{ lb} \quad Ans.$$

Solve all three equations for the unknowns, V, P, and H.

$$\therefore \quad H = P = 15 \text{ lb} \quad Ans.$$

Problem

An advertising sign is suspended as in Fig. 16. Its weight is 70.0 lb and the center of gravity is assumed to be at the center of the sign. What is the tension in the supporting cable? Calculate also the vertical and horizontal thrusts on the pivot P.

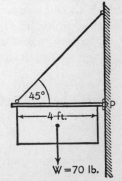

Fig. 16

Solution

Isolate the sign in the diagram (better here to redraw with forces indicated).

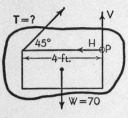

Fig. 17

Tabulate the force components and torques about P.

FORCE	X-COMP.	Y-COMP.	TORQUES ABOUT P
T	$T \cos 45°$	$T \sin 45°$	$-T \times 4 \sin 45°$
H	$-H$	0	0
V	0	$+V$	0
W	0	-70	$+70 \times 2$

Set up $\Sigma F_x = 0$; $\Sigma F_y = 0$

Set up $\Sigma L = 0$ about P

$$\left\{ \begin{array}{l} T \cos 45° - H = 0 \\ T \sin 45° + V - 70 = 0 \\ -4T \sin 45° + 140 = 0 \end{array} \right\} \begin{array}{l} \text{To be solved} \\ \text{by algebra.} \end{array}$$

Carry out solution. $4(.707)T = 140$

$$\therefore \quad T = \frac{140}{.707(4)} = 49.5 \text{ lb} \quad Ans.$$
$$H = .707(49.5) = 35.0 \text{ lb} \quad Ans.$$
$$V = 70 - .707(495) = 70 - 35 = 35.0 \text{ lb} \quad Ans.$$

Problems: Second Condition of Equilibrium

1. A 20 ft boom of a derrick is uniform and weighs 500 lb. At the end of the boom hangs a weight of 2000 lb. The boom is held at 30° above the horizontal by a cable at right angles to the boom at the outer end. Consider the hinge as the axis of rotation. (a) What is the magnitude of the torque produced by the weight of the boom? (b) What is the magnitude of the lever arm of the 2000 lb weight? (c) What is the magnitude of the torque produced by the hanging weight? (d) What is the magnitude of the torque produced by the cable on the boom? (e) What is the magnitude of the force exerted on boom by cable?

2. A 15 ft ladder is supported horizontally by vertical ropes attached to each end of the ladder. A man weighing 150 lb is sitting 3 ft from the left-hand end. The ladder weighs 30 lb. (a) Draw a

diagram showing all the forces acting on the ladder. (b) Set up a table showing the x- and y-components of the forces acting on the ladder and the torques about an axis through the right-hand end of the ladder. (c) What force is exerted by the left-hand rope? (d) What is the torque produced about the left end of the ladder due to the man's weight? (e) Would this torque be greater, less, or the same if one end of the ladder were lowered?

3. A 20 ft uniform plank weighing 100 lb leans against a smooth vertical wall. The plank is supported at the lower end by a horizontal platform and makes an angle of 60° with the vertical wall. (a) Draw a diagram showing all the forces acting on the plank. (b) What is the frictional force between the plank and the platform? (c) What is the torque about the lower end of the plank due to its own weight? (d) Would the torque be greater, less, or the same if the angle with the wall were increased? (e) If the plank is just on the verge of slipping, what is the coefficient of static friction between the plank and the platform?

4. A uniform yardstick (3 ft) weighing 1.00 lb is hung vertically from a nail through a hole in one end. The lower end is then pulled aside by a horizontal force F, of such magnitude that the stick makes an angle of 37° with the vertical. Find: (a) The vertical component of the force exerted by the nail on the stick. (b) The horizontal component of the force exerted by the nail on the stick. (c) The magnitude of F. (d) The magnitude of the resultant force at the nail.

5. A bamboo pole 10 ft long and weighing 16 lb rests against a frictionless wall with the foot of the pole 6 ft from the base of the wall and the top of the pole 8 ft above the ground. The center of gravity is 4 ft from the lower end. (Note that the sine of the angle between pole and ground is $\frac{8}{10}$ and the cosine of that angle is $\frac{6}{10}$.) (a) Draw a careful diagram of the pole, representing all forces acting on it by the proper vectors. (b) Determine the force exerted by the wall on the pole.

6. A uniform ladder 20 ft long and weighing 120 lb is carried by two men, one at one end and the other 4 ft from the other end. A weight of 40 lb is hung at the middle of the ladder. How much does each man lift?

7. A uniform rod 8 ft long and weighing 30 lb is pivoted 3 ft from one end. What weight suspended from the end of the short arm will balance a weight of 80 lb hung from the end of the long arm?

8. A 60 lb kitchen door 3 ft wide and 6 ft tall is supported by two hinges each 2 ft from the extremities of the door. If the lower

hinge supports all the weight, calculate the vertical and horizontal components of the forces acting on each hinge.

9. A uniform trap door in a floor is 4 ft square and is hinged along one edge. If the door weighs 100 lb and is opened upward by a vertical force so that it makes an angle of 60° with the horizontal, calculate the vertical and horizontal forces on the hinges.

10. A uniform derrick boom weighing 200 lb is hinged at one end to a vertical mast with respect to which it makes an angle of 60°, being held by a rope making an angle of 90° with its upper end and attached to the mast. What is the tension in the rope, and what is the resultant force acting on the hinge?

3

Kinematics and Dynamics of Translatory Motion

To study the motion of bodies not in equilibrium requires an appreciation of such kinematical concepts as position, displacement, velocity, and acceleration. These will now be considered with a view toward their application to problem situations. The continued theme is that a proper understanding of the definitions of the concepts makes problem solving a straightforward logical procedure.

Position is specified with respect to the origin of the mathematical x-, y-, z-coordinates (see Fig. 18).

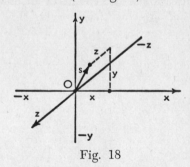

Fig. 18

Displacement (s) is a vector quantity defined 'as the change in position. In one dimension

$$s = x_2 - x_1$$

where x_2 represents a final position and x_1 represents an initial position. The distance a body travels in going from an initial to a final position is not necessarily the same as the displacement, since the latter is represented by a straight line from the initial position and pointing toward the final position, whereas the body may have moved over a circuitous path. In Fig. 18, displacement s is from the origin to position x, y, z.

Velocity (v) is defined as the time rate of displacement. The simple ratio of displacement over time is known as *average velocity* ($\bar{v}$), where the bar over a symbol stands for average. Thus

$$\bar{v} = \frac{s}{t} \text{ (a defining equation)} \quad \text{See page 2.}$$

The magnitude of velocity, which is a vector quantity, is called *speed*.

The average velocity over an infinitesimally small interval of time is referred to as *instantaneous velocity* (v). In mathematical language

$$v = \lim \frac{\Delta s}{\Delta t} \quad \text{as} \quad \Delta t \to 0 = \frac{ds}{dt} \text{ (using calculus notation)}$$

where Δs refers to a small increment of displacement and Δt refers to a correspondingly small increment of time.

Acceleration is defined as the time rate of change of velocity. Since the magnitude or the direction of velocity may change, or both of these quantities may change together, a vector operation is implied. If only the speed changes, i.e., if the direction is unchanged

$$a = \frac{v - v_0}{t} \text{ (a defining equation)} \quad = \frac{dv}{dt} = \frac{d^2s}{dt^2}$$
$$\text{(using calculus notation)}$$

where t represents the time during which the velocity changes and v_0 is the initial velocity. If, however, the direction only changes, as in uniform curvilinear motion (motion of a mathematical point in a circular path at constant speed), it can be shown that

$$a_c = \frac{v^2}{r} \text{ (a useful derived equation)}$$

where a_c is the central acceleration and r is the radius of the circular path. (The direction of the acceleration is perpendicular to that of the velocity at each instant and changes continuously in this case.)

Since the kinematic quantities displacement, velocity, and acceleration are vector quantities, they are subject to the convention of algebraic signs noted in the preceding chapter. Thus negative acceleration does not mean deceleration, but rather an acceleration opposed in direction to that which has been designated positive.

Types of Translatory Motion. The simplest conceivable motion is *uniform motion* characterized by the fact that the acceleration is zero, i.e., the velocity is constant. It does not change in magnitude or direction. Hence the motion is necessarily straight-line motion. Here $v = \bar{v}$, and since $\bar{v} = s/t$, it follows that

$$\boxed{s = \bar{v}t}$$ (a useful derived equation)

The next simplest type of motion is *uniformly accelerated linear motion,* characterized by the fact that the *acceleration is constant* both in magnitude and direction. This motion also is necessarily straight-line motion, but here the speed changes at a constant rate. Since by definition

$$a = \frac{v - v_0}{t}$$

it follows that $\boxed{v = v_0 + at}$

(a useful derived equation)

which means literally that the final velocity after an interval of time t is equal to the initial velocity v_0 plus the gain in velocity $a \times t$ during the interval. For this type of motion it is possible to write other equations as a result of algebraic manipulation of the defining equations. One of these is

$$\boxed{s = v_0 t + \tfrac{1}{2} at^2}$$ (a useful derived equation)

which means that the displacement is equal to the initial velocity multiplied by the time plus one-half the product of the acceleration and the square of the time. Another is

$$v^2 = v_0{}^2 + 2as$$ (a useful derived equation)

The type of uniformly accelerated motion for which the speed remains constant while the direction only of the velocity vector changes at a constant rate is called *uniform curvilinear motion*. It has already been mentioned. The significant derived equation for this motion is

$$a_c = \frac{v^2}{r}$$

Note! The preceding five boxed-in so-called useful derived equations are indeed the "formulas" by which all problems in linear kinematics are actually solved. Naturally they should be remembered (but not just memorized lest they not be recognized if different letters are used to represent the concepts). If the student derives them for himself several times he will find it difficult to forget them.

Free Fall. The motion of a freely falling body comes under the classification of uniformly accelerated linear motion since the direction is unchanging and since the speed is found to increase at a uniform rate under the pull of gravity alone. In this case the acceleration is designated as the *acceleration of gravity* and is usually represented by the symbol g. The value of g is approximately 32 ft/sec/sec (written 32 ft/sec^2) or 980 cm/sec/sec (980 cm/sec^2) or 9.8 m/sec^2. At a given place on the earth's surface the acceleration of gravity is constant.

Motions more complicated than those just listed are seldom encountered in first-year college Physics courses. Consequently problems illustrating kinematics will cover only the motions listed.

Problem Procedures

It should be realized that the concepts of motion are primarily displacement, velocity, and acceleration. Problems involving motion must deal with these concepts, so that basically their definitions are all-important. Indeed, all problems in kinematics must be limited to such situations as can be represented by relationships (formulas) that are readily derivable by mathematical logic from the defining equations. The best advice that can be given the beginner here is to have the definitions completely mastered. Care should be taken not to memorize symbols, since symbols alone are meaningless. They may even stand for different concepts at different times. The important thing is to understand the story which the symbols tell in each and every situation, realizing that by methods of mathematics it is possible to reason out the logic of every physical situation. If concepts and relationships are really understood then formulas will be remembered quite easily. In fact it will be difficult to forget them, because, as shorthand expressions, they are simply memory joggers themselves.

Probably the most difficulty will be encountered with *units*. In order to use defining equations most effectively, units must all be self-consistent. This is accomplished by reducing all data to basic units regardless of those used in the statement of the problem, i.e., feet, centimeters, or meters for length; slugs (explained later), grams, or kilograms for mass; and seconds for time in the English, cgs metric, or mks metric systems respectively. Such a procedure will result in a quantitative problem solution, coming out in the basic unit for the concept in question based on its defining equation.

An alternative procedure with units is to carry all units throughout the problem, treating them like algebraic quantities as regards factoring, multiplying, dividing, etc. Physics teachers are divided in their opinions regarding these two alternatives. One school of thought argues that by following the latter procedure the unit of an answer is worked out along with the concept in question, thus avoiding any mistake in units. The other school argues that carrying units throughout a problem is cumbersome and unnatural since it complicates enormously the arithmetic details and is rather arbitrary. The latter school's view is that relations exist between concepts and that there is merit in establishing a basic unit for each concept according to its defining equation (one unit in each system of units), and the three basic concepts length, mass, and time. Thus the arithmetic details involving units are confined to reducing the data to basic units before the physics of a problem is considered. Certainly this minimizes any possible confusion by separating the conversion of units from the analysis of conceptual relationships, which are two completely distinct operations.

In this text students will be encouraged to follow this procedure, i.e., first to reduce all data to basic units and then to forget the units until it is time to express the answer in that unit which is appropriate to the concept in question.

Problem

A ship sails due east covering a distance of 40 miles in 2 hours. What is its average speed? What is its average velocity?

Solution

Since this problem seems trivial it is to be noted that the reason for this feeling is simply that the defining equation for average speed $\bar{v} = s/t$ is generally well understood. The problem affords an opportunity to point out that if all concepts in physics were understood as well, all other problems in first-year college Physics would also seem trivial.

First step in solution:

Recall defining equation, or equations, of all important concepts involved. Here there is only one, namely

$$\bar{v} = \frac{s}{t}$$

Second step:

Manipulate the defining equations along with any relationships known to exist (previously derived) to get a mathematical expression for whatever concept is asked for in the problem. Here this step is unnecessary.

Third step:

Substitute numerical values and solve for the unknowns.

$\bar{v} = 40$ miles/2 hours $= 20$ miles/hour. This is the average speed. The average velocity is 20 miles/hour east.   *Ans.*

Note! Here it would have seemed unnatural to reduce the data to feet and seconds, but had the problem been complicated it would probably have been safer to do so even if the answer (29.3 ft/sec) is not so readily appreciated.

Problem

A car traveling in a straight line with initial speed of 20 miles per hour is accelerated uniformly to a speed of 40 miles per hour in 10 sec. What is the acceleration, and how far did the car travel during the interval?

Solution

The problem deals with uniformly accelerated linear motion which is charatcerized by $a =$ constant. The defining equation for acceleration is

$$a = \frac{v - v_0}{t}$$

If this is the only question, then numerical values may be substituted directly.

$$a = \frac{40 - 20}{10} \text{ miles/hr/sec}$$

$$\therefore \quad a = \tfrac{20}{10} = 2 \text{ miles/hr/sec} \quad \textit{Ans.}$$

To determine the displacement one recalls either that for this type of motion

$$s = v_0 t + \tfrac{1}{2}at^2$$

or that

$$s = \bar{v}t$$

At this point units must be considered, since time is specified in different units in the problem, and velocity is not expressed in basic units. Either all data must be reduced to basic units for the sake of consistency, or all units must be substituted along with numerical values into the appropriate equations. As noted earlier the former procedure is recommended in this book.

$$v_0 = 20 \text{ miles/hr} = \frac{20 \times 5280}{3600} \text{ ft/sec} = 29.3 \text{ ft/sec}$$

$$v = 40 \text{ miles/hr} = \qquad\qquad\qquad 58.6 \text{ ft/sec}$$

$$t = 10 \text{ sec}$$

Now $\quad a = \dfrac{v - v_0}{t} = \dfrac{29.3}{10} = 2.93$ ft/sec^2 *Ans.*

And $\quad s = v_0 t + \frac{1}{2}at^2$

$\qquad = 29.3(10) + \frac{1}{2}(2.93)(100)$

$\qquad = 293 + 146.5 = 439.4$ ft *Ans.*

Or $\quad s = \bar{v}t$ where $\bar{v} = \dfrac{v_0 + v}{2}$ (also a matter of definition)

$\therefore \quad s = \left(\dfrac{29.3 + 58.6}{2}\right) 10$

$\qquad = (43.95)(10) = 439.5$ ft *Ans.*

It should be noted that the first answer for the acceleration, namely 2 miles/hr/sec, is the same as

$$\dfrac{2 \times 5280}{3600} \text{ ft/sec}^2$$

$$= \dfrac{5280}{1800} = 2.93 \text{ ft/sec}^2, \text{ the second answer above}$$

Problem

How fast is a freely falling body moving at the end of 3 sec if it started from rest? How far has it fallen? (Neglect air resistance.) See Fig. 19.

Solution

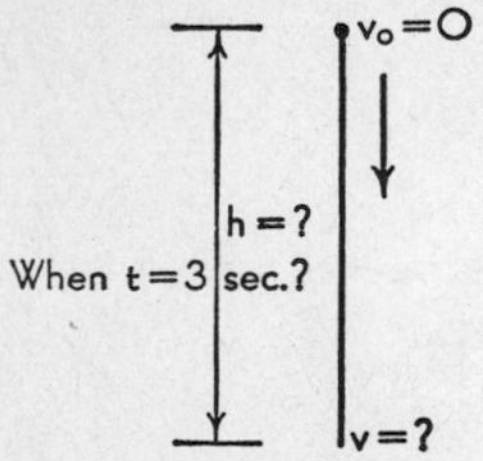

Recognize free fall as uniformly accelerated linear motion for which

$v_0 = 0$

$h = ?$

When $t = 3$ sec.?

$v = ?$

Fig. 19

$$v = v_0 - gt \quad s = v_0 t - \tfrac{1}{2}gt^2$$
$$\text{and} \quad v^2 = v_0^2 - 2gh$$

where the acceleration has the value g (32 ft/sec^2) and is negative.

$v = v_0 - gt = 0 - 32(3) = -96$ ft/sec

(The negative sign means the velocity is downward.) *Ans.*

Since $\quad v^2 = v_0^2 - 2gh$

$\quad h = \dfrac{v^2 - v_0^2}{2g} = \dfrac{(-96)^2 - 0}{2(32)} = \dfrac{9216}{64} = 144$ ft *Ans.*

Alternative procedure for this part of the problem:

$$s = v_0 t - \tfrac{1}{2} g t^2 = 0 - \tfrac{32}{2}(9) = 144 \text{ ft}$$

Problem

A body is thrown straight up at a speed of 16 ft/sec. How high does it rise? (Neglect air resistance.) See Fig. 20.

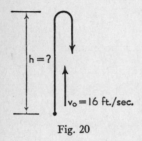

$h = ?$

$v_0 = 16$ ft./sec.

Fig. 20

Solution

Recognize the problem as one involving uniformly accelerated linear motion, where the acceleration is $-g$, and the initial velocity is positive and known. For such motion the three equations (assumed by now to be well known) are

$$\begin{cases} v = v_0 - gt \\ s = v_0 t - \tfrac{1}{2} g t^2 \\ v^2 = v_0{}^2 - 2gh \end{cases} \qquad \begin{aligned} v_0 &= 16 \text{ ft/sec} \\ g &= 32 \text{ ft/sec}^2 \end{aligned}$$

Realize the fact that the height to which the body rises is determined by the time required for the body to come to rest in its upward motion before it starts downward, i.e., that $v = 0$ at top of trajectory.

But $$v = v_0 - gt$$

$$\therefore \quad 0 = +16 - 32t$$
$$t = \tfrac{16}{32} = .5 \text{ seconds (time required to reach top)}$$

Since $$v^2 = v_0{}^2 - 2gh$$
$$0 = 16^2 - 2(32)h$$
$$h = \frac{16^2}{64} = \frac{256}{64} = 4 \text{ ft} \quad Ans.$$

Problem

A racing car traveling in a circular path maintains a constant speed of 120 miles per hour. If the radius of the circular track is 1000 ft, what if any acceleration does the center of gravity of the car exhibit?

Solution

Recognize the problem as an application of uniform curvilinear motion that does have an acceleration which is directed toward the center of the circle. Draw a diagram. See Fig. 21.
Recall the relationship for this acceleration.

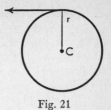

Fig. 21

$$a_c = \frac{v^2}{r} \qquad v = 120 \text{ miles/hr}$$

$$= \frac{120(5280)}{3600}$$
$$= 176 \text{ ft/sec}$$

and
$$r = 1000 \text{ ft}$$

$$\therefore \quad a_c = \frac{(176)^2}{1000} = \frac{30980}{1000} = 30.98 \text{ ft/sec}^2 \text{ toward center} \quad Ans.$$

Comment. This problem, like many in Physics, is almost trivial once it is recognized as falling into a type, i.e., recognized as exemplifying one of the motions studied. Whereas it appears to be merely a matter of substituting numerical values in a formula, which of course must be remembered, it is important for the student to visualize the concepts. This will help him to recognize the type of motion involved.

Problems: Kinematics of Translatory Motion

1. A motorist travels a distance of 10 miles in 20 min. During the trip he accelerates his car uniformly from 30 mph to 60 mph in 10 sec. (a) What is the average speed of the car for the entire trip in mph? (b) Is the speed constant for the 10 miles? (c) What is the acceleration of the car in the 10 sec interval? Express in ft/sec². (d) How far does the car go in the 10 sec interval?
2. A body has a displacement of 123 meters starting from rest. If its acceleration is 15.4 meter/sec² what is the time for this displacement?
3. How long will it take a body moving with an average speed of 15 ft/sec to go a distance of 6000 ft?
4. The runway on an airfield is 800 ft long. If an airplane, starting from rest, has to acquire a speed of 176 ft/sec in order to leave the ground, determine (a) the acceleration, assuming it to be constant; (b) the time necessary to acquire this speed.

5. A ball is dropped from the roof of a building and is timed until it reaches the ground. This time is 3 sec. How high is the building?

6. A 4 lb toy train travels around a circular track at a constant speed of 5 ft/sec. The radius of curvature of the track is 3 ft. What is the magnitude and direction of the train's acceleration?

7. Assuming the acceleration of gravity to be 9.8 meter/sec^2, how far will a freely falling body drop in 6 sec, starting from rest?

8. An object is thrown vertically upward with a velocity of 64 ft/sec. (a) How long will it take? (b) How high will it rise? (c) Where will it be 3 sec from the start?

9. A car traveling initially at 30 mph (44 ft/sec) is accelerated to 60 miles per hour in 10 sec. (a) What is the acceleration? (b) How far does it travel during the 10 sec?

10. An automobile traveling at constant speed rounds a curve at 48.8 km/hr. If the curve has a radius of 615 meters, what is the acceleration of the car?

11. A pebble is dropped into a well. If the splash is heard 3 sec after the pebble is released, how far is the water level below the point of release?

12. A block sliding down a frictionless plane travels 10 cm from rest in the first second, 30 additional cm in the second second, 50 additional cm in the third second, 70 additional cm in the fourth second, and 90 additional cm in the fifth second. What is the magnitude of its acceleration?

Linear Dynamics. *Dynamics,* in contrast with *Kinematics,* which deals only with motion in the abstract, deals with the motion and changes in motion of real bodies (bodies having mass) under the influence of forces.

Newton's Laws of Motion. The motions of bodies are found experimentally to be consistent with three laws discovered by Newton. These are:

I. A body at rest or in a state of uniform motion continues in that state of rest or uniform motion forever unless a resultant force acts on it to change its motion.

II. If a resultant force acts *on* a given body, an acceleration in the direction of the force and proportional in magnitude to it is acquired by the body. $F = MA$

III. To every force there is an equal and opposite reaction force.

These laws are not to be subjected to attempted proof by derivation from defining concepts. They are rather to be considered as laws of nature and the starting points in the derivations of other relationships, as well as the key to the solutions of many problems.

It should now be obvious to the student that the first law of Newton is the justification for the procedure followed in equilibrium problems, namely, to set the resultant of all the forces acting *on* a body equal to zero. The third law also explains the fact that for every force acting on the body, there is a corresponding one exerted *by* the body on something else.

The second law, however, is more general than the first law. It tells not only what the forces acting on a body add up to when the body is in equilibrium, but also what they add up to in general. In mathematical language the second law states that

$$\Sigma F = kma$$

or that the resultant force is proportional to the product of the mass of the body by its acceleration. Obviously if all the forces acting on a body of known mass can be evaluated, and if the value of the proportionality constant k is known, the acceleration of the body can be determined. If this proves to be such as to describe one of the motions studied in kinematics, then various aspects of the motion can be calculated from the appropriate kinematical formula. Thus the second law becomes the basis for the solution of many dynamical problems. It is a very powerful analytical tool.

Units. The proportionality factor k in the formula

$$F = kma$$

is associated with the units in which F, m, and a are expressed. By a proper choice of units for the concepts of force and mass, the value of k can be made unity since a basic unit of acceleration already exists (cm/sec^2 or ft/sec^2 or m/sec^2). Thereupon a unit of force equals a unit of mass times a unit of acceleration and

$$F = ma$$

In popular usage the "pound" is a unit often associated with mass and also with weight. Since weight is a force, however, it should not be expressed in mass units, or vice versa. The same difficulty is encountered with the "gram," which in popular usage sometimes is a unit of weight and sometimes a unit of mass.

Since the acceleration of gravity is the acceleration associated with the force of gravity (weight) on a body, it follows that

$$W = mg$$

Because of the lack of unanimity in the matter of force and mass units, and because the pound (or the gram) cannot be used for both force and mass in the same equation, students are required to be familiar with several systems of units, both English and metric. The so-called *British Engineering* system uses the *pound* as a unit of *force* and introduces the *slug* as the unit of *mass*. In the *English Absolute* system the *pound* is used as a unit of *mass* and the *poundal* is introduced as a unit of *force*. In the *Metric Absolute* system (cgs) the *gram* is retained as a unit of *mass* and the *dyne* is introduced as the unit of *force*. A fourth system, sometimes referred to as the *French Engineering* system, uses the *gram* (or the gram weight) as a unit of *force* and the unit of mass is nameless. (It is a mass whose weight is 980 g.)

In this book, as in most Physics courses with an engineering emphasis, when English units are used they will be limited to the British Engineering units, and when metric units are used they will be limited to Metric Absolute units, which include cgs (centimeter, gram, second) and also mks (meter, kilogram, second). See the next section of this text for a further discussion of this topic. Thus when British units are used

$$F \text{ (pounds)} = m \text{ (slugs)} \times a \text{ (ft/sec}^2\text{)}$$

The *slug* is a quantity of mass which weighs 32 lb (approximately). When metric units are used

$$F \text{ (dynes)} = m \text{ (grams)} \times a \text{ (cm/sec}^2\text{)}$$

$$F \text{ (newtons)} = m \text{ (kilograms)} \times a \text{ (m/sec}^2\text{)}$$

The dyne is the weight of an amount of mass equal to $\frac{1}{980}$ g, and the newton is the weight of $\frac{1}{9.8}$ kg, or approximately the weight of 100 g.

Accordingly, it is recommended in the solution of problems that masses be reduced to slugs, or grams, or kilograms, in the British or metric systems respectively, regardless of the data given in a problem, and that forces (including the force of gravity or weight) be reduced to pounds, or dynes, or newtons, in the British or metric systems respectively, before considering any formulas. To do this always *divide pounds by* 32 *to get slugs,* always *multiply grams by* 980 *to get dynes,* and always *multiply kilograms by* 9.8 *to get newtons.* This is very important in problem work since failure to do so will inevitably lead to unnecessary confusion.

Mks Units. A system of units known as mks is becoming increasingly popular in scientific work. In this system length is expressed in meters, mass in kilograms, and time in seconds. It is hoped that this system will eventually become universal and replace both the British and the cgs metric systems. This is far from the situation at present in practical problem work, as a result of which the mks system of units is to be treated as simply one of the systems described. In this system the *unit of force* is the *newton,* which is equivalent to 10^5 dynes, or approximately the weight of 100 g, or very approximately $\frac{1}{5}$ of a pound. It is the force necessary to give one kilogram an acceleration of one meter per second per second.

Problem Procedures

At this time the student should appreciate the reason for analyzing problems by considering all the forces that act on an isolated body. It is simply that Newton's Laws tell what happens to a body's motion under the influence of force. Therefore, certain specific steps are very logically suggested in the solution of all problems in dynamics.

1. Draw a diagram to represent the problem situation, labeling all quantities involved, and specifying the unknown as well as the known factors.

2. Isolate the body under consideration and indicate *all forces* acting *on* the body. Disregard the forces exerted *by* the body on everything else.

3. Tabulate the x- and the y-components of *all* the forces acting *on* the body.

4. Utilize Newton's second law, $F = ma$, but apply the law to the x- and the y-components separately, i.e.

set $$\Sigma F_x = ma_x$$

and $$\Sigma F_y = ma_y$$

5. Attempt to solve the two equations thus obtained for not more than two unknowns. If more than two quantities appear to be unknown, consider such relations as

$$W = mg$$

or the law of friction $$F = nN$$

as in statics or equilibrium problems.

Comment. If all the forces and the mass are known, this procedure will allow the acceleration to be calculated. If, however, the acceleration is known (given in the problem), then perhaps one or more of the forces acting on the body will be the unknowns.

6. If it is the acceleration which is calculated in the above, then the student should consider whether or not the problem fits one of the acceleration categories considered in the preceding kinematic study. If the category is recognized, then the equations for that motion (recall the bracketed formulas in kinematics) become applicable so that such factors as the velocity at any time, the initial velocity, the displacement, or the time come under consideration. Many a problem which at first seems quite involved yields a surprising amount of information under this straightforward and logical approach.

Caution. When substituting numerical values in the equations of step 4, be sure that the data have been reduced to basic units in the system used.

Problem

A 16 lb body is dragged along a horizontal surface by a horizontal force of 10 lb. The coefficient of friction between the body and the surface is .4. Starting from rest how far does the body move in 3 sec?

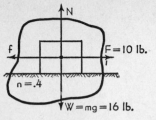

Fig. 22

Solution

Draw a diagram.

Isolate body and indicate all forces acting on body.

Tabulate x- and y-components of all forces *on* body and set $\Sigma F_x = ma_x$; $\Sigma F_y = ma_y$.

FORCE	X-COMP.	Y-COMP.
F	$+10$	0
W	0	-16
N	0	$+N$
f	$-f$	0

$$10 - f = ma_x$$
$$-16 + N = ma_y$$

But $\quad m = \frac{16}{32}$ slugs

and $\qquad f = .4N$ and $\quad a_y$ is obviously 0 (equilibrium)

$\qquad \therefore \quad N = 16$ lb

$$10 - .4(16) = \frac{16}{32}a_x = 10 - 6.4 = 3.6$$

$$\therefore \quad a = 3.6\left(\frac{32}{16}\right) = 7.2 \text{ ft/sec}^2$$

Note! $a =$ constant (7.2 ft/sec^2), whereupon the motion is uniformly accelerated linear motion for which

$$v = v_0 + at$$
$$s = v_0 t + \tfrac{1}{2}at^2$$
$$v^2 = v_0{}^2 + 2as$$

Moreover $v_0 = 0$ (given in problem).

$$\therefore \quad s = 0(3) + \tfrac{1}{2}(7.2)(3)^2 = 32.4 \text{ ft in 3 sec} \quad Ans.$$

Problem

An elevator is accelerated upward at 2 ft/sec². If the elevator weighs 500 lb, what is the tension in the supporting cable?

Solution

Diagram.
Isolate body and indicate forces on it.

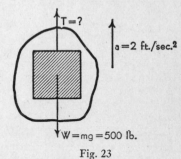

Fig. 23

Set $\Sigma F = ma$

$$T - mg = ma$$

Substitute values and solve for unknown.

$$T - 500 = \tfrac{500}{32}(2) = 31.2$$
$$\therefore \quad T = 500 + 31.2$$
$$= 531.2\ \text{lb}\quad Ans.$$

Problem

A 32 lb body is dragged along a practically frictionless horizontal surface by a 2 lb weight hanging from a cord which passes over a frictionless pulley so mounted that the cord pulls the body parallel to the surface. What acceleration is experienced by the body and its attached weight? What is the tension in the cord?

Solution

Diagram.

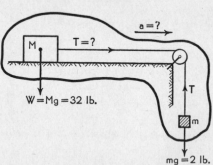

Fig. 24

Isolate the block, cord, and weight, since they all move as a unit.

Apply $F = ma$ in direction of motion
$$mg = (M + m)a$$

Substitute values and solve.

$$2 = \left(\tfrac{32}{32} + \tfrac{2}{32}\right)a$$
$$\therefore \quad a = \frac{2}{34/32} = 1.9 \text{ ft/sec}^2 \quad Ans.$$

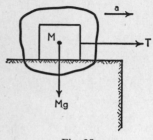

Fig. 25

Comment. Since T does not act externally on the combined system, but is an internal force, it does not enter into the above solution. To solve for T, one body only must be isolated, whereupon

$$T = \tfrac{32}{32}(1.9) = 1.9 \text{ lb} \quad Ans.$$

Problem

A 20 g projectile is fired vertically upward from a gun with an initial speed of 20,000 cm/sec. What is the acceleration of the projectile after leaving the gun? How long does it take to reach the top of the trajectory? Where is the projectile and how fast is it traveling 30 sec after leaving the gun? (Neglect air resistance.)

Solution

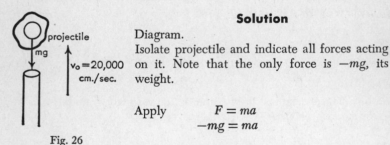

Fig. 26

Diagram.
Isolate projectile and indicate all forces acting on it. Note that the only force is $-mg$, its weight.

$$\text{Apply} \qquad F = ma$$
$$-mg = ma$$
$$\therefore \quad a = -g = -980 \text{ cm/sec}^2$$

(a points downward and is equal to the acceleration of gravity) *Ans.*

Therefore motion is uniformly accelerated linear motion since $a = $ constant.

$$\therefore \quad v = v_0 + at = 20,000 - 980t$$

At top of flight $v = 0$ $\therefore$ $0 = 20,000 - 980t$

$$\therefore \quad t = \frac{20{,}000}{980} = 20.4 \text{ sec to reach top} \quad Ans.$$

After 30 sec the displacement is given by

$$s = v_0 t + \tfrac{1}{2} a t^2 = 20{,}000(30) - \tfrac{1}{2}(980)900$$
$$= 600{,}000 - 441{,}000 = 159{,}000 \text{ cm}$$
$$\text{above starting point} \quad Ans.$$

And $\quad v = v_0 + at$
$$= 20{,}000 - 980(30) = -9{,}400 \text{ cm/sec (downward)} \quad Ans.$$

Problem

A body is whirled in a vertical circle by means of a cord attached at the center of the circle. At what minimum speed must it travel at the top of the path in order to keep the cord barely taut, i.e., just on the verge of collapse? Assume radius of circle to be 3 ft.

Solution

Diagram.
Isolate body.
Indicate forces on body. (Under conditions of problem there is no tension in cord.)

Set
$$\Sigma F = ma$$
$$-mg = ma \quad \therefore \quad a = -g$$

Fig. 27

At this stage recognize the motion as approximating uniform curvilinear motion, such that at the point in question:

$$a_c = -\frac{v^2}{r}$$
$$\therefore \quad -\frac{v^2}{r} = -g$$
$$\text{or} \quad v^2 = rg$$

(*Note!* Minus sign indicates that the direction of a is downward at top point of path.)

Substitute numerical values.

$$v^2 = 3(32)$$

$$v = \sqrt{3(32)} = \sqrt{96} = 9.8 \text{ ft/sec} \quad Ans.$$

Problems: Linear Dynamics

1. A body weighing 48 lb is accelerated 8 ft/sec^2. What is the resultant force in the direction of the acceleration?
2. An elevator which weighs 2000 lb is rising with an acceleration of 2 ft/sec^2. What is the tension in the cable which lifts it?
3. A resultant of 500 dynes acts toward the west on a mass of 10 g. What is the acceleration of the body?
4. Two weights of 8 lb and 16 lb respectively hang on the ends of a cord over a small frictionless pulley. If the system is released, what is the acceleration of the 16 lb weight?
5. A man weighs himself on spring scales while aboard an airliner. Although his actual weight is 160 lb, his apparent weight (scale reading) is 140 lb. The scales are assumed to be accurate. (a) What is the mass of the man? (b) Is. the vertical component of the acceleration up or down? (c) What is the vertical acceleration? (d) What is the apparent weight when the vertical velocity is constantly 40 ft/sec?
6. A sled acquires a certain speed at the foot of an incline so that it travels 92.5 m along a level surface before coming to rest. Assuming a coefficient of friction 0.2, what was the velocity at the foot of the incline?
7. A 29 kg body when placed on a spring balance in a moving elevator appears to weigh only 27.5 kg. What must be the acceleration (magnitude and direction) of the elevator?
8. A horizontal force acting on a 29 kg body moves it 9.25 m from rest in 3 sec on a level plane. If the coefficient of friction between body and plane is 0.1, what is the magnitude of the horizontal force?
9. A light cord passing over a vertical frictionless pulley of negligible mass has weights of 6 lb and 8 lb respectively attached to its ends. If released, what is the acceleration of the 6 lb body? the 8 lb body? What is the tension in the cord?
10. What tractive force is required to give a 3200 lb car an acceleration of 8.00 mph/sec?

Centripetal and Centrifugal Force. The problem of the body being whirled in a circular path was solved by a straightforward application of $F = ma$, where the acceleration a was recognized as characteristic of curvilinear motion. Since the product of mass and acceleration is dimensionally a force, there is obviously a force required to give a body curvilinear motion. It is called *centripetal force* and is expressed by the relation

$$F = ma_c = \frac{mv^2}{r}$$

It is directed toward the center of the circular path. The reaction to this force (Newton's third law) is called *centrifugal force,* which points away from the center of the path and is a force exerted by the body on something other than the body.

Problem

In a whirl-type washing machine the clothes are restrained to rotate in a circular path, but the water in the clothes is not so restrained because the confining drum is perforated. If 4 lb of clothes are assumed to be uniformly distributed about the interior of a drum of one foot radius, what force acting on the clothes (but not on the water droplets) tends to separate the water from the clothes at a rotating speed of one turn per second (equivalent to a linear speed of 6.28 ft/sec for the clothes in this case)?

Solution

Diagram.
Recognize that the resultant force is predominantly centripetal force.

$$F = \frac{mv^2}{r} = \frac{4}{32}\left(\frac{6.28}{1}\right)^2 = 4.93 \text{ lb} \quad Ans.$$

Comment. At top of path this force is decreased by 4 lb (weight) and at bottom it is increased by 4 lb.

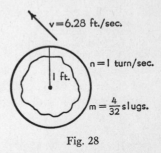

Fig. 28

Projectile Motion. An earlier problem illustrated what is called projectile motion although it dealt with a special case only, namely, that of a body projected vertically upward. In general, a projectile can be given an initial velocity in any direction. Therefore, projectile motion in general (not considering the effects of air resistance) is two-dimensional, i.e., the initial velocity has an x- as well as a y-component. It is to be noted that the x-component of the velocity is always constant (the x-motion is uniform linear motion), and the y-component of the motion is uniformly accelerated linear motion with the acceleration of gravity.

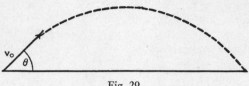

Fig. 29

$$a_x = 0; \; v_x = v_{0x} = v_0 \cos \theta; \; x = v_x t$$
$$a_y = -g; \; v_y = v_{0y} - gt = v_0 \sin \theta - gt$$
and $$y = v_{0y}t - \tfrac{1}{2}gt^2 = v_0 \sin \theta \, t - \tfrac{1}{2}gt^2$$

Problem Procedures

Treat the two components of the motion separately. The key to the solution of the y-motion is the realization that at the top of the trajectory the value of $v_y = 0$.

But in general $v_y = v_0 \sin \theta - gt$.

Therefore the time required to reach the top is

$$t = \frac{v_0 \sin \theta}{g}$$

This value of t when substituted into the equation for y (vertical displacement) yields the height of the trajectory. This same value of t when substituted into the equation for x (horizontal displacement) yields the half-value of the range.

Problem

A mortar shell is fired at an angle of 60° with the horizontal. If its initial velocity is 1000 ft/sec in this direction, how high does it rise and what is its range? (Neglect air resistance.)

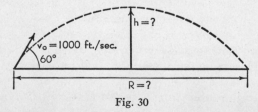

Fig. 30

Solution

Diagram.
Either recognize the problem as a projectile problem, or isolate the shell and indicate the forces acting on it to show that while in flight only the force of gravity (exclusive of air resistance) acts on it. Thus the vertical motion is characterized by

$$a_y = -g$$

and the horizontal motion is constant, i.e., $a_x = 0$.

$$\therefore \quad v_x = v_0 \cos \theta \quad \text{and} \quad v_y = v_0 \sin \theta - gt$$

Also $\quad x = v_0 \cos \theta\, t \quad$ and $\quad y = v_0 \sin \theta\, t - \frac{1}{2}gt^2$

First, calculate the time required to reach the top, i.e., to make

$$v_y = 0 = v_0 \sin \theta - gt$$
$$\therefore \quad t = \frac{v_0 \sin \theta}{g}$$

Substituting values $\quad t = \dfrac{1000 \sin 60°}{32} = \dfrac{866}{32} = 27 \text{ sec (approx)}$

In this time it will rise to a height given by

$$y = v_0 \sin \theta\, t - \tfrac{1}{2}gt^2 = 1000(.866)(27) - \tfrac{1}{2}(32)(27)^2$$
$$= 23,400 - 11,700 = 11,700 \text{ ft}$$
$$\therefore \quad h = 11,700 \text{ ft} \quad Ans.$$

During this same time the horizontal displacement is

$$x = v_0 \cos \theta\, t = 1000(.5)(27) = 13,500$$

The range R is equal to twice this value of x.

$$\therefore \quad R = 2(13,500) = 27,000 \text{ ft} \quad Ans.$$

Momentum and Impulse of Force. *Momentum* is a concept defined as the product of mass multiplied by velocity.

$$\text{Momentum} = mv \text{ (a vector quantity)}$$

Momentum is important because of the conservation principle which can easily be derived by considering Newton's second and third laws together, and which states that in a collision of two or more bodies in an isolated system the total momentum of the system remains unchanged. This means that *after* a collision the sum of the momenta of all bodies concerned is equal to the sum of the momenta of all bodies concerned *before* the collision if no external forces act on the entire system.

The change in momentum of a given body can be shown to be equal to the force acting on the body multiplied by the time during which the momentum changed

$$mv - (mv)_0 = ft$$

The product ft is called *impulse of force.*

Problem

A 4000 lb automobile traveling at 30 mph overtakes and collides with a 2000 lb car traveling in the same direction at 15 mph. Their bumpers lock and they then move as one body. How fast does the combined system move, disregarding the frictional drag on each car?

Solution

Recognize this as a situation involving the principle of the conservation of momentum.

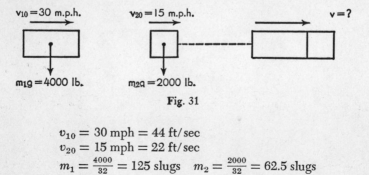

Fig. 31

$$v_{10} = 30 \text{ mph} = 44 \text{ ft/sec}$$
$$v_{20} = 15 \text{ mph} = 22 \text{ ft/sec}$$
$$m_1 = \frac{4000}{32} = 125 \text{ slugs} \quad m_2 = \frac{2000}{32} = 62.5 \text{ slugs}$$

Here $m_1v_{10} + m_2v_{20} = (m_1 + m_2)v$.
Solve for v and substitute values.

$$v = \frac{m_1 v_{10} + m_2 v_{20}}{m_1 + m_2} = \frac{125(44) + 62.5(22)}{187.5}$$
$$= 37 \text{ ft/sec (approx)} \quad Ans.$$

Problem

A boy standing on very slippery ice (assume it to be frictionless) throws a 5 lb object away from him in a horizontal direction with a speed of 2 ft/sec. If he weighs 80 lb, how fast will he start to move in the opposite direction?

Solution

Recognize this as a situation involving conservation of momentum, realizing that whereas the original momentum is zero, the final momentum must also be zero. Because of the vector nature of momentum the combined momentum of boy and object after the incident can be zero while the momentum of each need not be, providing they point in opposite directions.

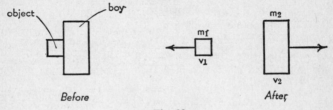

Before   After

Fig. 32

$$(m_1 + m_2)0 = -m_1 v_1 + m_2 v_2$$
$$\therefore \quad v_2 = \frac{m_1 v_1}{m_2} = \frac{\frac{5}{32}(2)}{\frac{80}{32}} = \frac{1}{8} \text{ ft/sec} \quad Ans.$$

Problem

A rifle has a mass of 25 kg and fires a bullet of 10 g mass at a muzzle velocity of 9×10^4 cm/sec. What is the recoil velocity of the rifle?

Solution

$$M = 25 \text{ kg} \quad mv = MV$$
$$m = 10 \times 10^{-3} \text{ kg}$$
$$v = 9 \times 10^2 \text{ m/sec}$$
$$V = ?$$

$$V = \frac{mv}{M} = \frac{10^{-2} \times 9 \times 10^2}{2.5 \times 10} = 3.6 \times 10^{-1}$$
$$= .36 \text{ m/sec} \quad Ans.$$

Problems: Centripetal Force, Projectiles, Momentum

1. A boy on a bicycle pedals around a circle of 40 ft diameter at a speed of 20 ft/sec. The weight of boy and bicycle is 160 lb. (a) What is the centripetal acceleration of the boy and bicycle? (b) What is the centripetal force exerted on the bicycle? (c) What is the friction force between tires and road that is directed toward the center of the circle? (d) What is the minimum value of the coefficient of friction between tires and road that will prevent slipping?

2. A fly rests on the edge of a 12 in. ($r = 6$ inches) phonograph record. The record rotates so that the speed of the fly is 2 ft/sec. The weight of the fly is .0032 lb. (a) What is the mass of the fly? (b) What is the centripetal acceleration of the fly? (c) What is the centripetal force on the fly? (d) What is the minimum coefficient of friction that will keep the fly from slipping?

3. A stone which weighs 0.5 lb is tied to a string 2 ft long. If the breaking strength of the string is 15 lb: (a) Determine the maximum speed with which the stone can be whirled in a horizontal circle. (b) Determine the magnitude of the acceleration of the stone when it is at its breaking speed. (c) What is the magnitude of the momentum of the stone when the string is under maximum tension, that is, when the stone has the speed found in part (a)?

4. A 60 lb hunting dog leaps from a 40 lb canoe which also contains a 160 lb hunter. The canoe containing dog and hunter is originally at rest. The speed of the dog as he leaves the canoe is 20 ft/sec. (a) What is the velocity of the canoe just after the dog leaves? (b) What is the momentum of the hunter and canoe just after the dog leaves?

5. A car weighing 3200 lb is moving with a speed of 60 mph. What is its momentum? (60 mph = 88 ft/sec)

6. A rifle bullet is shot into the air at an angle of 37° with the horizontal at a speed of 1200 ft/sec. Neglect air resistance. (sin 37° = .6; cos 37° = .8; tan 37° = .75.) (a) What is the vertical component of the initial velocity? (b) What is the horizontal component of the initial velocity? (c) How long is the bullet in the air? (d) How far horizontally does the bullet go?

7. A 3200 lb car rounds a curve of 400 ft radius at a speed of 30 mph. How large must the force of friction be between tires and pavement to prevent the car from skidding?

8. An 8 lb body suspended from a 6 ft cord is swung around a vertical axis like a conical pendulum, making 60 rpm. What is the tension in the cord and the radius of the horizontal circle in which the body revolves?

9. A small block is projected horizontally off the edge of a table 4 ft high with a velocity of 12 ft/sec. How far from the base of the table and with what velocity (magnitude and direction) does it strike the floor?

10. A 4 oz shell is fired horizontally from a 16 lb rifle. If the shell is given a velocity of 900 ft/sec, what is the velocity of recoil of the gun?

11. A 10 g bullet is fired into a 2 kg wooden block, giving the block a velocity of 20 cm/sec. With what velocity was the bullet fired?

4

Work and Energy

Although Newton's laws of motion provide a very powerful method of attack on problems in mechanics, they do not give the only one. In problems where time is not specifically mentioned, the concept of energy provides an approach because of an all-important conservation principle, called the *conservation of energy*.

Energy is a concept defined as the capacity to do work, where work is a technical concept not to be confused with its popular reference to muscular fatigue. If while a body is displaced a force acts on it, and if the force, or a component of it, acts in the direction of the displacement, the product of this force, or its component, and the displacement is called *work*. For a constant force f acting through a displacement s

$$W = fs$$

For a variable force, using the calculus notation:

$$W = \int F \, ds$$

Energy and work are obviously measured in the same units. These are foot-pounds in the English system, and dyne-centimeters or ergs in the cgs system, and newton-meters or joules in the mks system. One joule is equivalent to 10^7 ergs.

There are several kinds of energy, including potential energy and kinetic energy. *Potential energy* is energy due to *position,* whereas *kinetic energy* is energy due to *motion*. More precisely, kinetic energy is defined by the relation:

$$KE = \tfrac{1}{2}mv^2$$

and potential energy is energy stored in a body by virtue of work having been done on it by a so-called *conservative force,* or one whose value does not depend upon velocity, but rather upon position. Examples of such a force include the force of gravity and the force of a spring. Therefore there is potential energy due to elevation

$$PE = mgh$$

and a spring has potential energy

$$PE = \tfrac{1}{2}kx$$

when stretched (or compressed, as the case may be) by an amount x. (More about springs later.)

Conservation of Energy. In all mechanical processes the sum of the potential and kinetic energy remains constant for the system concerned. If there is work done against friction in the process, this is not included in the above-mentioned summation because the force of friction is not a conservative, but rather a *dissipative,* force, which therefore cannot be associated with potential energy. Another statement of the principle of conservation of energy is that, excluding such forces as friction, the gain in kinetic energy of a system must equal the loss of potential energy in the same system during any process.

Power. The rate at which work is done is called *power.* In English units it is measured in ft-lb per second. Thus 550 ft-lb/sec, or 33,000 ft-lb/min, represents one horsepower. In cgs units power is measured in ergs/sec and the mks unit of power is the joule/sec or the watt. One thousand watts make a kilowatt and 746 watts equal one horsepower.

$$p = \frac{W}{t}$$

Problem

A sled starting from rest at the top of an incline slides with a minimum of friction (assume it to be frictionless) from an elevation of 60 ft to the base. How fast is it traveling when it reaches the base?

Solution

This problem can be solved by the method of force analysis described earlier: setting $\Sigma F = ma$, finding the acceleration to be constant, and calculating the velocity at the end of the displacement along the incline. Since, however, the time is not specifically mentioned in the problem, the energy method is suggested.

Diagram.

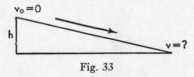

Fig. 33

Note! The total energy $(PE + KE)$ at the top equals the total energy $(PE + KE)$ at the base.

AT TOP		AT BASE
$PE + KE$	$=$	$PE + KE$
$mgh + 0$	$=$	$0 + \frac{1}{2}mv^2$

$$\therefore \quad mgh = \frac{1}{2}mv^2$$
$$v^2 = 2gh \qquad \therefore \quad v = \sqrt{2gh}$$

Substitute values and solve.

$$v = \sqrt{2(32)(60)} = \sqrt{3840} = 62 \text{ ft/sec (approx) } Ans.$$

Problem

A simple pendulum consists of a small object (a so-called bob) hanging from a relatively long cord whose weight is negligible with respect to the bob. The to-and-fro motion of this bob in a vertical plane is called pendulum motion. If the cord is 3 ft long and the suspended bob is drawn back so as to allow the cord to make an angle of 10° with the vertical before being released, calculate the speed of the bob as it passes through its lowest position.

Solution

This problem also can be solved by force analysis, but it lends itself most readily to a solution by the energy method.

Diagram.

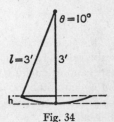

TOP OF SWING		BOTTOM OF SWING
$PE + KE$	$=$	$PE + KE$
$mgh + 0$	$=$	$0 + \frac{1}{2}mv^2$
$\therefore \quad v = \sqrt{2gh}$		

To determine h, note that it is $l - l \cos \theta$

Fig. 34

$$= 3 - 3 \cos 10° = 3 - 3(.985)$$
$$= 3 - 2.96 = .04 \text{ ft}$$
$$v = \sqrt{2(32)(.04)} = \sqrt{2.56} = 1.6 \text{ ft/sec} \quad Ans.$$

Problem

The ballistic pendulum is a device which makes possible the determination of the velocity of a projectile which is fired into a block supported as a simple pendulum. A bullet of 20 g mass is fired into a block of 500 g so supported from a 100 cm cord, causing it to swing through an angle of 10° with the vertical. Calculate the velocity of the bullet.

Solution

Diagram.

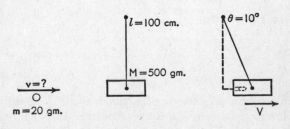

Fig. 35

Utilizing the principle of the conservation of momentum and the principle of the conservation of energy

$$mv = (M + m)V$$
and $$V = \sqrt{2gh}$$
where $$h = l - l\cos\theta$$
$$\therefore \quad v = \frac{M + m}{m}\sqrt{2gl(1 - \cos\theta)}$$

Substituting values

$$v = \frac{500 + 20}{20}\sqrt{2(980)100(1 - .985)}$$
$$= 26\sqrt{1960(1.5)}$$
$$\therefore \quad v = 26\sqrt{2940} = 26(54)\,(\text{approx}) = 1400 \text{ cm/sec (approx)} \quad \textit{Ans.}$$

Problems: Work and Energy

1. A .15 kg ball starting from rest falls freely from a height of 35 m above the ground. (a) What is the velocity just before it hits the ground? (b) How long did it take to reach the ground? (c) What is the potential energy of the ball at the start with respect to the ground? (d) How much work would have to be done on the ball to restore it to its original position? (e) What is the kinetic energy of the ball just before it hits the ground?

2. A car at rest at the top of a hill has 96,600 ft-lb of potential energy. If it weighs 3200 lb, what is the velocity of the car at the bottom of the hill where the potential energy is zero? The friction may be neglected.

3. A 160 lb man skis down a hill, the top of which is 100 ft above the surrounding plain. At the bottom, his speed is 60 ft/sec. The slope is 350 ft long. (a) What is the potential energy of the man when at the top with respect to the bottom? (b) What is the kinetic energy of the man at the bottom of the slope? (c) What energy is expended in overcoming friction between skis and snow? (d) What is the frictional force between skis and snow?

4. A baseball weighing approximately $\frac{1}{3}$ lb is hit in a line drive. Its speed after being hit is 150 ft/sec. (a) What is the magnitude of the momentum of the ball just after it is hit? (b) What is the kinetic energy of the ball just after it is hit? (c) The ball is caught by the third baseman who moves his hands back 1 ft while stopping the ball. What is the average force exerted on the ball by the player's glove? Assume no loss in kinetic energy during flight. (d) What becomes of the kinetic energy of the ball when it is caught?

5. A boy standing on a bridge 200 ft above a river throws a $\frac{1}{4}$ lb stone straight downward with a velocity of 50 ft/sec. Neglect air resistance. (a) With what speed will the stone strike the water? (b) How long will it take to descend? (c) What is the potential energy of the stone just as it leaves the boy's hand (with respect to the surface of the water)? (d) What is the total energy of the stone just as it leaves the boy's hand?

6. A 320 lb crate is dragged along a horizontal floor by a rope making an angle of 30° with the floor. The tension in the rope is 40 lb. How much work is done in dragging the crate 20 ft?

7. A 72.9 kg man climbs a flight of stairs 3.7 m high in 6 sec. How much work does he do, and how much power does he expend?

8. A 2 lb projectile is fired vertically upward with a speed of 600 ft/sec. (a) What is its potential energy 3 sec later? (b) What is its kinetic energy at the highest point of its flight? (c) How much potential energy does it have at this point? (d) What is its kinetic energy as it returns to the starting point?

9. A cake of ice moving along a perfectly smooth horizontal surface at a speed of 8 ft/sec comes to a ramp sloping upward at an angle of 30° with the horizontal. If the ramp also is frictionless, how far along it will the ice travel before coming to rest?

10. How much energy is required to stop a 3200 lb car traveling at 60 mph? If this is accomplished in 10 sec, what is the average deceleration and how much power is expended?

Machines. Since work is a compound concept, the product of force and displacement, it is obvious that a given amount of work can be expressed in a variety of ways. A small force multiplied by a large displacement can represent the same amount of work as a large force multiplied by a small displacement.

$$fS = Fs$$

This is the same as
$$\frac{F}{f} = \frac{S}{s}$$

which suggests that in doing a given amount of work, a force advantage becomes possible if a corresponding displacement disadvantage is tolerated. A *machine* is a device for multiplying force (at the expense of displacement) or of changing the direction of a force. It should be noted that no machine does more work than is done on it. Actually the friction introduced by the working parts of machines results in all machines doing somewhat less work than is done on them. Thus the idea of efficiency enters into machine considerations.

$$Eff = \frac{\text{Output Work}}{\text{Input Work}}$$

Designating F/f as the *actual mechanical advantage* (MA_A) of a machine and S/s as the *theoretical mechanical advantage* (MA_T) of a machine

$$Eff = \frac{MA_A}{MA_T}$$

Problem Procedures

Problems involving machines usually deal with the one or the other mechanical advantage. The theoretical advantage can generally be determined by considering the geometry of the device. For example, a movable pulley mounted as in the first diagram of Fig. 36 requires that the length of the rope pulled at f must

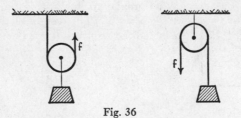

Fig. 36

be twice the distance the load is lifted. Hence the theoretical advantage of such an arrangement is 2. On the other hand, if the arrangement is that of the second diagram in Fig. 36, there is no sacrifice of displacement and hence no theoretical gain. The theoretical advantage is simply 1.

In the case of the lever, the wheel and axle, the inclined plane, the worm gear, etc., as well as the pulley, the procedure is to assume a fictitious displacement of the load, and then to figure out the displacement associated with the applied force by considering the dimensions, or the number of gear teeth, or the geometry in general. For the lever, the theoretical advantage is the ratio of the lever arms. For the block and tackle, it is the number of ropes attached to the movable pulleys. For the inclined plane, it is the ratio of the length of the incline to the elevation. For the wheel and axle, it is the ratio of the radius of the large wheel to the radius of the axle, for obviously in one complete revolution (see Fig. 37) the displacement of a point on the wheel

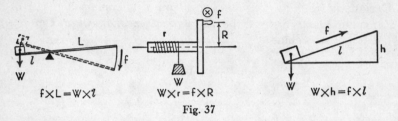

$$f \times L = W \times l \qquad W \times r = f \times R \qquad W \times h = f \times l$$

Fig. 37

where the force is applied is R/r times the displacement of a point on the axle, where the load acts.

Problem

A cylinder is rolled up an incline by a tape arranged as in Fig. 38. What minimum force f is required if the angle $\theta = 30°$? The weight of the cylinder is 2 lb and the elevation $h = 2$ ft.

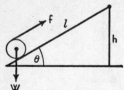

Fig. 38

Solution

Recognize this as a machine problem in which a weight W is lifted a distance h by a device which has mechanical advantage. The device is essentially two machines, the one an inclined plane and the other a movable pulley. The inclined plane has a theoretical advantage given by

$$\frac{l}{h} = \frac{l}{l \sin \theta} = \frac{1}{\sin \theta}$$

Superimposed on this advantage is the theoretical advantage of the movable pulley, which is 2.

$\therefore$ The combined theoretical advantage is $\dfrac{2}{\sin \theta}$

Thus the minimum force (the force required if the frictional drag is a minimum or when the efficiency is unity) is found to be

$$f = \frac{W}{2/\sin \theta} = \frac{W}{2} \sin \theta$$

Substituting numerical values:

$$f = \tfrac{2}{2}(.5) = \tfrac{1}{2} \text{ lb} \quad Ans.$$

Comment. This answer is consistent with the conclusion that the work done on the machine equals the work done by the machine.

$$f(2l) = Wh$$
$$\tfrac{1}{2}(8) = 2(2)$$
$$4 = 4$$

All machines, no matter how complicated, are essentially combinations of the several basic machines such as the pulley, the inclined plane, the lever, and the wheel and axle, for which the theoretical advantage in each case is readily determined.

Problems: Machines

1. A chain hoist lifts a 1 ton load (2000 lb) by the application of a 20 lb force on the loop of the chain. What is the actual mechanical advantage of this machine?
2. A jack lifts an 800 lb load 6 in. when a force of 50 lb is applied to the end of the 2 ft handle. What is the actual mechanical advantage of the jack?
3. A steel safe weighing 1 ton is to be loaded onto a truck body 5 ft above the ground by sliding it up a plank 20 ft long. If it takes 200 lb to overcome friction, what is the least force necessary to push the safe up the plank?
4. By the use of a wheel and axle a 400 lb concrete block is raised by

80 lb applied to the rim of the wheel. If the radius of the axle is 4 in., and that of the wheel is 2 ft, (a) what is the actual mechanical advantage, (b) the theoretical mechanical advantage, and (c) the efficiency?

5. By the use of an inclined ramp whose rise is 3 ft in 10 ft along the incline, a 200 lb chest is dragged from a basement to the first floor of a building whose floors are 9 ft apart. What *minimum* force is required? Assuming an efficiency of 50%, what force would be required?

6. If a 6 ft bar is used by a person to lift a 150 lb rock and a fulcrum can be located 1 ft from the rock end of the bar, what minimum force need be applied to the bar by the person, and where should it be applied?

7. In splitting a tree with a wedge whose taper is 1 in. each 6 in. of length, what splitting force is applied to a log by a 50 lb hammer blow upon the wedge?

8. By the use of a block and tackle a 400 lb piano can be lifted by a force of 90 lb. If each block has two pulleys, show by a diagram how they must be connected.

5

Rotary Motion

The study of rotary motion is simplified by emphasizing certain analogies with translatory motion. By defining rotational concepts so as to make each one correspond to a translatory concept, the mathematical relationships between the rotary ones become identical to those between the translatory concepts. It is customary to represent the rotational concepts by Greek letters.

Rotary Kinematics. Position on a line (X-axis), indicated by the symbol x, suggests the concept of *angular position* measured with respect to a given direction and represented by the symbol θ. (See Fig. 39.)

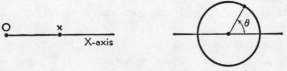

Fig. 39

The change in angular position is *angular displacement* ϕ.

$$\phi = \theta_2 - \theta_1 \backsimeq s = x_2 - x_1$$

The symbol $\backsimeq$ stands for "is analogous or corresponds to." Similarly *angular velocity* is the time rate of angular displacement. Average angular velocity $\bar{\omega}$ is given by

$$\bar{\omega} = \frac{\phi}{t} \backsimeq \bar{v} = \frac{s}{t}$$

$$\omega = \frac{d\phi}{dt} \backsimeq v = \frac{ds}{dt}$$

Furthermore *angular acceleration* is the time rate of change of angular velocity and is defined by the relationship

$$\alpha = \frac{\omega - \omega_0}{t} = \frac{d\omega}{dt} = \frac{d^2\phi}{dt^2} \approx a = \frac{v - v_0}{t} = \frac{dv}{dt} = \frac{d^2s}{dt^2}$$

It follows that for constant, or uniform, angular motion

$$\begin{cases} \alpha = 0 \quad \therefore \quad \omega = \text{const.} = \bar{\omega} \quad \phi = \bar{\omega}t \text{ just as} \\ a = 0 \quad \therefore \quad v = \text{const.} = \bar{v} \quad s = \bar{v}t \end{cases}$$

Also, for uniformly accelerated rotary motion

$$\begin{cases} \alpha = \text{const.} \quad \omega = \omega_0 + \alpha t \quad \phi = \omega_0 t + \frac{1}{2}\alpha t^2 \quad \omega^2 = \omega_0^2 + 2\alpha\phi \\ \quad \text{just as} \\ a = \text{const.} \quad v = v_0 + at \quad s = v_0 t + \frac{1}{2}at^2 \quad v^2 = v_0^2 + 2as \end{cases}$$

Thus it is seen that rotary kinematics is merely an extension of linear kinematics. Although rotary motion may seem more complicated than translatory motion to the beginner, its formulas are mathematically exactly the same but expressed in the appropriate Greek symbols instead of in English symbols. The important thing is to know what each concept means in order properly to interpret the symbolism.

Note! The importance of a thorough understanding of the material in Chapter 3 should now be obvious. What could have been a very difficult topic (Rotary Motion) is made relatively simple. Again the value of studying the concepts of Physics in logical sequence is emphasized.

Problem Procedures

Problems in rotary motion follow the same pattern as those in translatory motion. The given motion is first classified, and then the appropriate relationships (formulas) become applicable. Here it is interesting to note certain basic connections between linear and rotary motion

$$s = r\phi \qquad v = r\omega \qquad a = r\alpha$$

The unit of angular measure used in scientific work is the

radian. It is a dimensionless quantity, being the ratio of the circumference of a circle to the radius. In effect it is an angle equal to $1/2\pi$th of a whole circle. The unit for angular velocity is radian/second, and the unit for angular acceleration is radian/sec².

Problem

A flywheel rotating at 1800 rpm is allowed to come to rest, which it does in 10 min. Assuming that the speed decreases uniformly calculate (a) the angular acceleration and (b) the number of revolutions made in coming to rest.

Fig. 40

Solution

Diagram.

Recognize motion as uniformly accelerated rotary motion for which the following hold.

$\omega = 1800$ rpm
$t = 10$ min
$\alpha = ?$
$\phi = ?$

$$\begin{cases} \omega = \omega_0 + \alpha t \\ \phi = \omega_0 t + \frac{1}{2}\alpha t^2 \\ \omega^2 = \omega_0{}^2 + 2\alpha\phi \end{cases}$$

Convert data to proper units.

$$\omega_0 = 1800 \text{ rpm} = \tfrac{1800}{60} \text{ rps} = \tfrac{1800}{60}(2\pi)$$
$$= 60\pi \text{ rad/sec}$$

Also $t = 10$ min $= 600$ sec

Consider the first equation in above list expressed as $\alpha = \dfrac{\omega - \omega_0}{t}$.

$$\alpha = \frac{0 - 60\pi}{600} = \frac{-\pi}{10} \text{ rad/sec}^2 \text{ (opposite to the direction of motion)} \quad Ans.$$

Substituting in the third equation above expressed as $\phi = \dfrac{\omega^2 - \omega_0{}^2}{2\alpha}$

$$\phi = \frac{0 - (60\pi)^2}{2(-\pi/10)} = + \frac{3600\pi^2}{\pi/5} = 18,000\pi \text{ radians}$$
$$= \frac{18,000\pi}{2\pi} = 9000 \text{ rev} \quad Ans.$$

Problems: Rotary Kinematics

1. A wheel with a radius of 2 ft is rotating with a constant angular acceleration of 3 rad/sec². At $t = 0$ the angular speed is 10 rev/sec. (a) What is the linear speed of a point on the rim at $t = 0$? (b) Through how many radians will the wheel have turned in the time from $t = 0$ to $t = 5$ sec?

2. A wheel starting from rest acquires an angular velocity of 30 rad/sec with a uniform angular acceleration of .6 rad/sec². Through what angle (in radians) has it turned?

3. A wheel 2 ft in radius has zero angular acceleration and an angular velocity at time $t = 0$ of 5 rad/sec. (a) What is the angular velocity at $t = 10$ sec? (b) How many revolutions per minute does it make at $t = 0$? (c) With what linear speed does a point on the rim travel at $t = 0$?

4. A flywheel rotating at 1600 rpm is disengaged from its power supply and comes to rest in 16 sec. What is the angular deceleration, and how many revolutions does the flywheel make in coming to rest?

5. A rotor starting from rest with a constant acceleration acquires an angular speed of 300 rpm in 15 sec. Calculate (a) the angular acceleration, (b) the angular velocity at the end of the 10th second.

6. A wheel 2 ft in radius is rotating about its axle with a speed of 4 turns per second. (a) What is the angular velocity in radians per second? (b) What is the linear speed of a point on the rim? (c) If it starts to slow down and comes to rest in 30 sec, what is its average acceleration?

Rotary Dynamics. Pursuing the analogies further, the concept of moment of inertia in rotary motion corresponds to the concept of mass in translatory motion. Since torque or moment of force (L) corresponds to force (f), and angular acceleration (α) corresponds to linear acceleration (a)

$$f = ma \quad \text{suggests by analogy} \quad L = I\alpha,$$

where I stands for moment of inertia.

Although moment of inertia I in rotation is analogous to mass m in translation, it is a more complicated concept than mass because it incorporates not only the concept of mass but also the distribution of mass with respect to the axis of rotation.

$$I = \Sigma mr^2 = \int r^2 dm = \frac{L}{\alpha}$$

Since there are any number of possible axes of rotation for a body, there is no such thing as *the* moment of inertia of a body. When expressed with respect to an axis passing through the center of mass, the designation is I_0. Values of I_0 for three common objects are given as follows:

Hoop	$I_0 = mr^2$
Disk (Axis $\perp$ to face)	$I_0 = \dfrac{mr^2}{2}$
Sphere	$I_0 = \frac{2}{5}mr^2$

Values for other uniform homogeneous bodies are to be found in engineering tables.

It can be shown that with respect to an axis parallel to one through the center of mass but displaced from it by a distance h

$$I = I_0 + mh^2$$

The proper units for moment of inertia are slug ft^2 in the British Engineering system, gram cm^2 in cgs units, and kilogram m^2 in mks units.

Radius of Gyration. Because of the nature of moment of inertia, it is possible to express it as the product of a mass and the square of a radial distance. In the case of a hoop the radial distance, or distance from the axis at which all the mass can be considered to be concentrated, is the radius itself, but for other objects it is the square root of the ratio of the moment of inertia to the mass. It is designated as the radius of gyration (ρ).

$$\rho = \sqrt{\frac{I}{m}} \quad \text{or} \quad I = m\rho^2$$

Energy of Rotation. A rotating body has kinetic energy due to motion. It is given by the expression

$$KE_{\text{rot}} = \tfrac{1}{2}I\omega^2 \approx KE_{\text{trans}} = \tfrac{1}{2}mv^2$$

The law of the conservation of energy includes rotational kinetic energy as well as translational kinetic energy.

Problem

A uniform disk 1 ft in diameter and weighing 8 lb is rotated about an axle perpendicular to its face. What force applied tangentially to the rim of the disk will produce an angular acceleration of 2 rad/sec²? What kinetic energy will the disk have 4 sec after starting from rest?

Solution

Diagram.

$$d = 1 \text{ ft}; r = \tfrac{1}{2} \text{ ft}$$
$$m = 8 \text{ lb}$$

Recognize motion as pure rotation for which the torque L about the axis O is given by

$$= \tfrac{8}{32} \text{ slugs}$$
$$\alpha = 2 \text{ rad/sec}^2$$

$$L = fr = I_0 \alpha$$

And: $I_0 = \tfrac{1}{2}mr^2$ (disk)

$$\therefore f = \frac{\tfrac{1}{2}mr^2 \alpha}{r} = \frac{1}{2} \cdot \frac{8 \cdot 1 \cdot 2}{32 \cdot 2} = \frac{1}{8} \text{ lb} \quad Ans.$$

Fig. 41

Since $\alpha = 2$ rad/sec², the motion is uniformly accelerated rotary motion for which the following applies:

$$\omega = \omega_0 + \alpha t$$
$$= 0 + 2(4) = 8 \text{ rad/sec}$$

And:
$$KE = \tfrac{1}{2}I\omega^2$$
$$= \tfrac{1}{2}(\tfrac{1}{2}mr^2)\omega^2$$
$$= \tfrac{1}{2}(\tfrac{1}{2} \cdot \tfrac{8}{32} \cdot \tfrac{1}{4})(8)^2 = 1 \text{ ft-lb} \quad Ans.$$

Problem Procedures in Dynamics
(Including Linear and Rotary Motion)

For problems in dynamics generally, the following approach is recommended. It will be recognized by the student as the same approach suggested earlier for problems in statics; then extended first to cover linear dynamics; and finally extended to cover rotary dynamics, since every motion of a body can be resolved into translatory motion of the center of mass, and rotary motion about the center of the mass of the body.

1. Draw a diagram, labeling all quantities.
2. Isolate the body under consideration.
3. Indicate all forces acting on the isolated body.
4. Tabulate separately the x-components and the y-components of all the forces acting on the body.
5. Set $\Sigma F_x = ma_x$, and $\Sigma F_y = ma_y$
6. If only two quantities are unknown, solve the two equations thus obtained. If more than two quantities are unknown, establish a possible axis of rotation.
7. Tabulate (in a third column of the same table as in step 4) the values of the torques developed by the forces with respect to the chosen axis. Each of these will be a product of a force multiplied by a lever arm, the latter being the perpendicular distance from the axis to the line of action of the force. Torques tending to produce clockwise rotation are negative, and those tending to produce counterclockwise rotation are positive.
8. Set $\Sigma L = I\alpha$, noting that I stands for the moment of inertia of the body in question with respect to the chosen axis. If this axis does not pass through the center of mass, it should be recalled that

$$I = I_0 + mh^2$$

where h is the distance between an axis passing through the center of mass and any axis parallel to it.
9. Realize that the equations for the translational part of the motion and the rotational part may be related by such equations as

$$a = r\alpha \quad \text{or} \quad v = r\omega$$

10. Solve as simultaneous equations all the relationships that result from the preceding steps.
11. Reduce all data to basic units for the system of units used in the problem; then substitute numerical values and solve.
12. If the above does not yield as many equations as there are

unknowns in the problem, consider a different axis about which to sum torques. Different lever arms will result. This step can be repeated indefinitely.

Comment. Although the above procedure may seem lengthy and unnecessarily cumbersome, there is satisfaction in the fact that an answer can usually be obtained. It completely eliminates the frustration that some students develop when the conclusion is reached that one has to have special intuition, even genius, to solve such problems.

Of course many problems will not require the "full treatment." In some cases certain steps will be unnecessary, but the general procedure should be mastered. For those problems in which time is not mentioned explicitly, this procedure may well be short-circuited by resorting to the energy method described earlier.

Problem

A string is wrapped around a uniform homogeneous 3 lb cylinder with a 6 in. radius. The free end is attached to the ceiling from which the cylinder is then allowed to fall (as in Fig. 42), starting from rest. As the string unwraps, the cylinder revolves. What is the linear acceleration of the center of mass? What is the linear velocity, and how fast is the cylinder revolving after a drop of 6 ft? What is the tension in the cord?

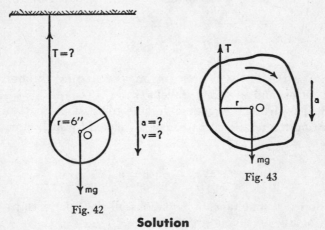

Fig. 42

Fig. 43

Solution

Diagram (already given).

Isolate the cylinder, and indicate forces acting *on* it.

There is no need to tabulate x- and y-components since they are all up or down forces.

Set $$\Sigma F = ma$$

(Note that in such problems it is convenient to take the direction of motion as positive instead of following literally the rule that up is positive and down is negative.)

$$mg - T = ma \qquad \therefore \quad a = g - \frac{T}{m}$$

Since the above yields insufficient information, consider rotation.

Consider torques about the center of mass 0, and set $\Sigma L = I\alpha$ and $Tr = I_0\alpha$.

(*Note!* Clockwise rotation corresponds to downward motion, already assumed to have the positive direction.)

But I_0 for a cylinder $= \frac{1}{2}mr^2$ (cylinder is an elongated disk)

and in this problem $$a = r\alpha$$

$$\therefore \quad T = \frac{1}{2}\frac{mr^2}{r}\frac{a}{r} = \frac{ma}{2}$$

whereupon $a = \dfrac{-ma}{2m} + g = \dfrac{-a}{2} + g$

$\therefore \quad 3a = 2g$

$\qquad a = \frac{2}{3}g = \frac{2}{3}(32) = 21.3$ ft/sec^2 (downward) *Ans.*

Now since $a = $ constant $(\frac{2}{3}g)$, the linear motion is uniformly accelerated, such that

$$v^2 = v_0^2 + 2as, \text{ where } s \text{ becomes the drop } h$$
$$= 0 + 2(\tfrac{2}{3}g)(6) = \tfrac{24}{3}g = 8g = 256$$
$$\therefore \quad v = \sqrt{256} = 16 \text{ ft/sec} \quad Ans.$$

The angular velocity $\omega = \dfrac{v}{r} = \dfrac{16}{\frac{1}{2}} = 32$ rad/sec *Ans.*

Also $T = m(g - a) = m \left(g - \frac{2}{3}g \right) = \frac{mg}{3} = \frac{3}{3} = 1$ lb $\quad$ *Ans.*

Solution by Energy Method

The loss in potential energy corresponding to a drop h is equal to the kinetic energy gained (translational plus rotational).

$$mgh = \tfrac{1}{2}mv^2 + \tfrac{1}{2}I_0\omega^2$$

But: $\quad I_0 = \tfrac{1}{2}mr^2$

and: $\quad r^2\omega^2 = v^2$

$\therefore \quad mgh = \tfrac{1}{2}mv^2 + \tfrac{1}{4}mv^2 = \tfrac{3}{4}mv^2$

$\qquad v^2 = \tfrac{4}{3}gh$

$\therefore \quad v = \sqrt{\tfrac{4}{3}(32)(6)} = 16$ ft/sec $\quad$ (Same answer as before.)

Also: $\quad v^2 = v_0^2 + 2ah$

But: $\quad v_0 = 0$, and $v^2 = \tfrac{4}{3}gh \quad$ (See above.)

$\therefore \quad a = \dfrac{v^2}{2h} = \dfrac{4}{6}g = \dfrac{2}{3}g$

$\qquad = 21.3$ ft/sec^2 (downward) $\quad$ (Same answer as before.)

Note again that when the energy method is applicable it yields the answer more directly than does the force-torque analysis.

Solution by Method of Instantaneous Axis

It is noted that with respect to the point where the cord just leaves the cylinder, a point that can be considered an instantaneous axis, the motion is entirely rotational such that

$$mgr = I\alpha, \quad \text{where} \quad I = I_0 + mr^2$$
$$mgr = \left(\frac{mr^2}{2} + mr^2 \right) \frac{a}{r} = \frac{3}{2}mra$$
$$\therefore \quad a = \tfrac{2}{3}g \quad \text{(Same answer as before.)}$$

Problem

A flywheel has a mass of 30 kg and a radius of gyration of 2 m. If it turns at a rate of 2.4 rev/sec, what is its rotational kinetic energy?

Solution

$$m = 30 \text{ kg}$$
$$\rho = 2 \text{ m}$$
$$\omega = 2.4 \text{ rps} = 2.4 \times 2\pi = 4.8\pi \text{ rad/sec}$$
$$KE = \tfrac{1}{2}I\omega^2 \quad \text{where} \quad I = m\rho^2$$
$$= \tfrac{1}{2}m\rho^2 \, (4.8\pi)^2$$
$$= \frac{30}{2} \times 2^2 \times (4.8)^2\pi^2$$
$$= 13{,}850 \text{ joules} \quad Ans.$$

Problems: Rotary Dynamics

1. A 12 lb weight hangs by a cord wrapped around a 2 ft radius drum which is free to turn about a fixed axis O. The angular acceleration of the drum is 3 rad/sec². (a) What is the acceleration of the 12 lb weight? (b) What is the tension in the cord? (c) What is the torque acting on the drum? (d) What is the moment of inertia of the drum?

2. A flywheel has a moment of inertia of 12 slug ft² and a weight of 100 lb. It is accelerating at a constant rate of 2 rad/sec². The mass may be assumed to be concentrated in the rim. (a) Calculate the radius of gyration. (b) Calculate the diameter. (c) Calculate the resultant torque acting on the wheel. (d) Starting from rest, how long did it take the wheel to reach a speed of 10 rad/sec?

3. A wheel has a radius of 12 cm, a mass of 2000 g, and a moment of inertia of 10^5 g cm². It rolls along a flat surface with a linear speed of 15 cm/sec. (a) With what linear speed does the axle move? (b) What is the angular speed of the axle? (c) What is the kinetic energy of translation? (d) What is the kinetic energy of rotation about the axle? (e) What is the linear momentum? (f) What is the angular momentum about the axle?

4. A man stands on a rotating platform, with frictionless bearings, which is turning at the rate of 6 rad/sec. The kinetic energy of rotation of man plus platform is 18 ft-lb. (a) What is the moment of inertia of the man plus platform? (b) The man stretches his arms out so that the angular speed becomes 3 rad/sec. What is the angular momentum of man plus platform under these conditions?

5. A flywheel has a moment of inertia of 100 slug ft² and a radius of gyration of 4 ft. It is subject to a torque of 8 lb ft. (a) Calculate the mass. (b) Calculate the weight. (c) Calculate the angular

acceleration. (d) How long will it take to produce a change in angular velocity of 10 rad/sec?

6. A uniform disk having a mass of 2 kg and a radius of 10^{-1} m rotates about its geometric axis at a rate of 200 rpm. What constant force applied tangentially at the rim will stop it in 40 sec?

7. A uniform cylindrical drum weighing 15 lb and having a radius of 6 in. is rotating about its geometrical axis at a speed of 2 rps. (a) What is its moment of inertia about this axis? (b) What is its radius of gyration? (c) If a tangential drag force of 1 lb is applied at the rim, what acceleration is produced? (d) How long will it take to come to rest? (e) How many revolutions will it make in coming to rest?

8. How far will a hoop of 6 in. radius roll up a plane making a 30° angle with the horizontal if its linear speed at the bottom is 5 ft/sec?

9. A 60 lb automobile tire-wheel combination has a moment of inertia of 3 slug ft². (a) What is its radius of gyration? (b) If a braking torque of 2 lb ft is applied, what angular acceleration is produced?

10. A 2 lb hoop of 6 in. radius rolls along the floor with an angular velocity of 90 turns per minute. What is its total kinetic energy?

6

Periodic Motions

Next in order of difficulty beyond the types of motion already discussed come the periodic motions. These are characterized by the fact that the motion follows a cyclic repetitive pattern. The length of time required for a complete cycle is known as the *period* (T). The reciprocal of the period is the *frequency* (n)

$$n = \frac{1}{T}$$

Simple Harmonic Motion. The most important, and very common, periodic motion is simple harmonic motion. It is the type of motion exemplified by the bouncing up and down of a body suspended from a vertical stretched cylindrical spring. It is characterized mathematically by the fact that the acceleration is proportional to the displacement (x) from the equilibrium position (the motion is a to-and-fro linear motion about a point of equilibrium) and directed toward it.

$$a = -cx$$

where c is a constant.

To derive equations for displacement, velocity, and acceleration (the common kinematical concepts usually associated with motion), reference is made to a circle because, interestingly enough, uniform circular motion ($a_c = v^2/r$) when projected on a diameter gives a motion along the diameter which is one and the same as simple harmonic motion. This means that in order to analyze simple harmonic motion a so-called *circle of reference*

81

may be considered. It is as if every to-and-fro linear (S.H.M.) motion is a uniform circular motion viewed edgewise. The *amplitude* (the maximum displacement measured from the equilibrium position) is the radius of the circle, as in Fig. 44.

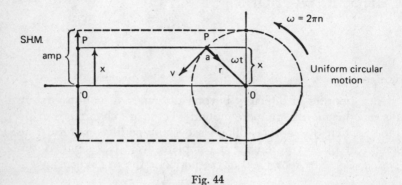

Fig. 44

Measured from the vertical, since a vertical S.H.M. is illustrated, the angle made by the radius from point P' to the center of the circle is ωt. Hence the *displacement* x in S.H.M. is the same as $r \cos \omega t$ in the circular motion. Moreover the *velocity* of P in S.H.M. is the projection of $v = r\omega$ along the vertical diameter of the circular motion, and the acceleration of P in S.H.M. is the same as the projection of $a_c = v^2/r$ along the diameter of the circular motion at the point P'.

Thus

$$x_{\text{S.H.M.}} = r \cos \omega t$$
$$v_{\text{S.H.M.}} = -2\pi n r \sin \omega t$$
$$a_{\text{S.H.M.}} = \frac{-v^2}{r} \cos \omega t = -4\pi^2 n^2 x = -cx$$

where c is a constant of proportionality. The last of the preceding equations when expressed in terms of n gives

$$n = \frac{1}{T} = \frac{1}{2\pi} \sqrt{-\frac{a}{x}}$$

These are the equations of simple harmonic motion that are available as formulas for problem solving. Incidentally, the minus signs mean only that the vector quantities concerned (velocity and acceleration) point in the direction opposite to the displacement.

Problem Procedure

If, after analyzing the forces acting on a body, the acceleration proves to be of the form

$$a = -cx$$

then the motion involved is simple harmonic motion, and the above equations become applicable.

Problem

One end of a fingernail file is clamped in a vise and the other end is given a to-and-fro vibration. The motion of the free end is approximately S.H.M. If the frequency is 10 vibrations per second and the amplitude is 4 millimeters, what is the velocity when the displacement of the free end is 2 millimeters?

Solution

The problem states that the motion is S.H.M. Therefore

$$v = 2\pi nr \sin \omega t \quad \text{(disregarding direction)}$$

But ωt is the angle associated with the imaginary circle of reference such that $r \cos \omega t = x$.

$$\therefore \quad \cos \omega t = \frac{x}{r} = \frac{2}{4} = .5$$
$$\therefore \quad \omega t = 60° \quad \text{and} \quad \sin \omega t = .866$$

Consequently

$$v = 2\pi (10)(4)(.866)$$
$$= 69.3\pi = 218 \text{ mm/sec} \quad Ans.$$

Incidentally, the acceleration is found to be

$$a = -4\pi^2 n^2 x$$
$$= -4\pi^2 \cdot 10^2 \cdot 2 = 8000 \text{ mm/sec}^2$$
$$= 800 \text{ cm/sec}^2 \text{ at the point in question}$$

Pendulum Motion. Insofar as simple pendulum motion is approximately linear (if the angle of swing is limited to approximately 5° either side of the vertical), it approximates simple harmonic motion. It can be shown that

$$T = \frac{1}{n} = 2\pi \sqrt{\frac{l}{g}}$$

where l represents length. This is often referred to as the simple pendulum formula.

In the case of the physical pendulum (not a mass bob at the end of a weightless cord, but a physical object swinging on a pivot), the period is given by the relation

$$T = 2\pi \sqrt{\frac{I}{mgh}} = 2\pi \sqrt{\frac{I/mh}{g}}$$

where I is the moment of inertia with respect to the axis in question, and h is the distance between the axis and the center of mass. The quantity I/mh is referred to as the length of the equivalent simple pendulum.

Problem

What is the acceleration of gravity at a place where the period of a simple pendulum 100 cm long is exactly 2 sec?

Solution

$$T = 2\pi \sqrt{\frac{l}{g}} \qquad 2^2 = \frac{(2\pi)^2(100)}{g}$$

$$\therefore \quad g = \frac{\cancel{4} \cdot \pi^2 \cdot 100}{\cancel{4}} = 1000 \text{ cm/sec}^2 \quad Ans.$$

Note! The accepted value of the acceleration of gravity is around 980 cm/sec² at standard latitude and sea level.

Problems: Simple Harmonic Motion

1. A weight hanging from the lower end of a vertical coiled spring bobs up and down with a frequency of 10 vibrations per sec and an amplitude of 5 in. How long does it take for the weight to make

one complete round trip? (b) What is the displacement when the velocity is zero? (c) What is the direction of the acceleration when the weight is at the top of the path?

2. A body of mass 3000 g executes simple harmonic motion with a frequency of 3 vib/sec and an amplitude of 6 cm. (a) What is the radius of the reference circle? (b) What is the maximum velocity of the body? (c) At what point of its path is the body when its velocity has its maximum value? (d) What is the maximum acceleration of the body? (e) What is the acceleration of the body when its displacement is 1 cm? (f) At what point in its path is the body when its acceleration is a minimum? (g) What resultant force acts on the body when its displacement is 6 cm?

3. A body vibrates in simple harmonic motion with an amplitude of 20 cm and a frequency of $\frac{1}{2}$ vib/sec. (a) What is the velocity at the extreme ends of the path? (b) What is the maximum velocity of the body? (c) What is the magnitude of the acceleration at the extreme ends of the path? (d) What is the acceleration of the body when it is at center point of the vibration?

4. A body executing simple harmonic motion has a maximum speed of 300 cm/sec and an amplitude of 200 cm. Calculate the period of the motion.

5. A seconds pendulum (period of 2 sec) has a length of 100 cm. What is the acceleration of gravity at the locality in question?

6. An 8 lb body is attached to a vertical coiled spring which gives it simple harmonic motion with a frequency of one complete vibration in 2 sec and an amplitude of 1 ft. What is the maximum kinetic energy of the body?

7. A body of 200 g performs simple harmonic motion with an amplitude of 3 cm. If its maximum acceleration is 2 cm/sec², (a) what is its period? (b) What is its kinetic energy when it passes through the equilibrium position? (c) What is its potential energy at the extreme end of its path?

8. A body of mass 2 kg attached to a suspended spring vibrates up and down with simple harmonic motion between positions A and B about an equilibrium position C. The amplitude is .1 m and the frequency is 5 vib/sec. (a) What are the velocity and the acceleration at a point midway between C and A? (b) What maximum force acts on the body by the spring?

Part II.

Properties of Matter and Sound

7

Statics of Elasticity
and Mechanics of Fluids

The preceding section (Part I) of this text has dealt with external forces only acting on so-called rigid bodies, and the motions of such bodies. No consideration has been given to internal forces or to the behavior of extensible bodies. A study of the mechanics of deformable bodies involves a consideration of various properties of matter.

Elasticity. All matter is characterized by a relative tendency to recover from distortion, such as a change of shape or volume or both, resulting from the application of force. This property, by virtue of which matter tends to resist and recover from deformation, is called *elasticity*. Among other things it affords a means of classifying matter as either *solid* or *fluid*, depending upon the extent to which it is displayed. Solid matter is *rigid*, meaning that the distance between any two mass points in a rigid body always remains constant. Matter is *fluid* when it is not rigid. The distinction between the solid state and the fluid state is more or less relative. Fluids are subdivided into two groups, *liquids* and *gases*, depending upon whether or not a free surface is displayed. Thus a study of elasticity leads logically to a study of liquid phenomena and gaseous phenomena, as well as to a study of the characteristics of solid matter.

Elastic Concepts. When a body is distorted, strains develop. *Strain* is defined as fractional deformation. Three types of strain are listed as follows:

1. Simple one-dimensional stretch or compression is expressible as $\Delta x/x$, a dimensionless ratio since the units of Δx are the same as those of x itself. See Fig. 45a.

2. Over-all fractional change in volume is expressible as $\Delta V/V$, also a dimensionless ratio. See Fig. 45b.
3. A shearing or twisting type of strain is represented by an angle as in Fig. 45c.

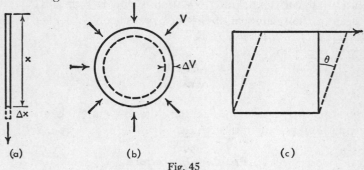

(a)　　　　　　(b)　　　　　　(c)

Fig. 45

When a body is strained, stresses are developed. *Stress* is defined as the restoring force per unit cross-sectional area developed internally.

Hooke's Law. It is a law of experience that *within the elastic limit* stress is always proportional to strain. The proportionality constant is called the *modulus of elasticity*. Referring to the three cases of strain listed above, it is evident that three different moduli are involved. These are called *Young's Modulus, bulk modulus,* and the *coefficient of rigidity,* respectively. They are expressed as follows:

$$Y \text{ (Young's Modulus)} = \frac{\dfrac{F}{A}}{\dfrac{\Delta x}{x}} \quad \text{(lb/sq ft or dynes/cm}^2 \text{ or newtons/meter}^2\text{)}$$

$$B \text{ (Bulk modulus)} \quad = \frac{\dfrac{F}{A}}{\dfrac{\Delta V}{V}} = \frac{p}{\dfrac{\Delta V}{V}} \quad \text{(Same units as for } Y.)$$

$$n \text{ (Coef. of rigidity)} \quad = \frac{\dfrac{F}{A}}{\theta} \quad \text{(Same units as above.)}$$

Comment. In the case of B above, the force is everywhere perpendicular to A. In such circumstances the ratio F/A is defined as *hydrostatic pressure p*. In the case of n, however, the force is tangential to A, and the ratio F/A is *not* pressure. In all cases the modulus is expressed in units of force per unit area, since the denominator of the defining equation is dimensionless.

Finally, the reciprocal of the bulk modulus B is defined as the *coefficient of compressibility C*.

$$C = \frac{1}{B}$$

Values of the various moduli for different materials are to be found in engineering tables.

Problem Procedures

Problems in elasticity are usually simple applications of the defining equations. Therefore, they can hardly involve more than mere substitutions of numerical values in these equations and the solving for the unknown quantities. This means that once these definitions are *understood*, problems involving them reduce to exercises in simple algebra and arithmetic where the only likely difficulties will be those associated with the conversion of units if the data are not expressed in basic units in the system used.

Problem

If Young's Modulus for steel is 19×10^{11} dynes/cm^2, how much force will be required to stretch a sample of wire 1 sq mm in cross section by 10% of its original length?

Solution

The problem is recognized as one obviously involving Young's Modulus, the defining equation for which is

$$Y = \frac{\frac{F}{A}}{\frac{\Delta x}{x}}$$

In this case $\Delta x/x = 1/10$ and Y is given as 19×10^{11} dynes/cm^2. Moreover $A = 1$ mm$^2 = .01$ cm^2.

$$\therefore \quad \frac{F}{A} = Y\frac{\Delta x}{x} \quad \text{and} \quad F = YA\frac{\Delta x}{x}$$

$$F = 19 \times 10^{11} \times 10^{-2} \times 10^{-1} = 19 \times 10^{8} \text{ dynes} \quad \textit{Ans.}$$

Problem

A 10 kg mass resting on a table is attached to a steel wire 5 m long which hangs from a hook in the ceiling, exactly 3 m above the mass. If the support is removed, the wire is momentarily stretched and the mass commences to oscillate up and down. Calculate the frequency of the oscillation if the cross-sectional area of the wire is 1 sq mm and the value of Young's Modulus is taken to be 19×10^{11} dynes/cm^2.

Solution

This problem is recognized as a problem combining the concept of Young's Modulus with a motion suspected to be approximately simple harmonic motion.

$$Y = \frac{\frac{F}{A}}{\frac{\Delta x}{x}} \quad \therefore \quad F = Y\frac{\Delta x}{x}A$$

But $F = ma$ (Newton's second law, where a = acceleration).

$$\therefore \quad a = \frac{F}{m} = \left(\frac{Y}{m}\frac{A}{x}\right)\Delta x \quad \text{whence} \quad a \sim \Delta x$$

$$\therefore \quad \text{Motion approximates S.H.M.}$$

But for S.H.M. $a = -4\pi^2 n^2\,\Delta x$ (Δx in this case)

$$\therefore \quad 4\pi^2 n^2\,\cancel{\Delta x} = \frac{Y}{m}\frac{A}{x}\,\cancel{\Delta x} \text{ (neglecting negative signs)}$$

$$\therefore \quad n^2 = \frac{YA}{4\pi^2 mx} \quad \text{and} \quad n = \frac{1}{2\pi}\sqrt{\frac{YA}{mx}}$$

Substituting numerical values after reducing the data to consistent units (here the cgs system is used)

$$n = \frac{1}{2\pi}\sqrt{\frac{19 \times 10^{11} \times .01}{10000 \times 500}} = \frac{1}{2\pi}\sqrt{\frac{19 \times 10^{9}}{5 \times 10^{6}}}$$

$$= \frac{1}{2\pi} \sqrt{3800} = \frac{62}{6.28} = \text{approximately 10 vib/sec} \quad Ans.$$

Problems: Elasticity

1. A wire is stretched elastically by a force F so that its length is increased by .2 inches. The length of the wire is 20 ft and it has a cross-sectional area of .04 sq inches. Young's Modulus is 5×10^7 lb/in.^{2}. (a) What is the stress? (b) What is the strain?
2. A wire is stretched elastically by a force F so that its length is increased by .2 in. The length of the wire is 10 ft and it has a cross-sectional area of .03 sq in. Young's Modulus is 3×10^7 lb/sq in. (a) What is the stress? (b) What is the force F?
3. Assuming Young's Modulus for steel to be 19×10^{11} dynes/cm^2, what will be the stretch of a steel wire 400 cm long and 2 sq mm in cross section when a stretching force of 6 kg is applied?
4. If the bulk modulus of brass is 8.5 lb/sq in., how much pressure will be required to compress a small brass sphere by one part in a hundred?
5. A cube of soft rubber 3.0 cm on an edge has two parallel and opposite forces of 300,000 dynes applied to opposite faces. If the shear is .10, what is the shear modulus of the material?

Concepts Applicable to Matter in Bulk. In studying properties of matter it is necessary to use language which differentiates a given body from bulk matter. One can properly refer to the mass, or even the weight, of a particular body, but in attempting to convey the idea that inherently a substance like lead is more massive than a substance like aluminum, the concept of mass is inadequate.

Density. *Density* is defined as mass per unit volume.

$$d = \frac{m}{V}$$

This is expressed in grams per cubic centimeter or kilograms per cubic meter when metric units are used. In English units it might seem natural to express density in pounds per cubic foot, until one recalls that the pound is a unit of force (weight) and not mass. Literally, then, in accordance with the earlier commitment to use British Engineering Units whenever English units are used, density must be expressed in slugs per cubic foot. There is much confusion over this point, and the student is warned to consider it carefully in all problem work. The natural desire in practical work to refer to the weight per unit volume leads to another concept called *weight density* $d_{wt} = W/V$. This must not be confused with true density in the analyses of problem situations.

Specific Gravity. It is often convenient to compare the density of a substance to that of a so-called standard substance, usually water. *Specific gravity* of a substance is the ratio of the density of that substance to the density of water. It is obviously a dimensionless quantity.

$$S_g = \frac{d_s}{d_w}$$

Here an advantage of cgs metric units is found. Since the density of water is for all practical purposes 1 g/cc, the numerical value of d_s in metric units is the same as that of the specific gravity of a given substance. This is not true in British Engineering Units. Because one cubic ft of water weighs 62.4 lb, its density is $\frac{62.4}{32.2} =$ 1.94 slugs/cu ft. (Although g has been rounded to 32 in this text, its more precise value of 32.2 should be used here.)

For problem purposes it should be realized that the weight of a body can be expressed as

$$W = mg = dVg = S_g d_w Vg \ (\text{pounds, dynes, or newtons})$$

Concepts and Principles Applicable to Fluids at Rest. Fluids include liquids and gases, the former being more or less incompressible (constant density), whereas the latter are quite compressible. Thus the concept of fluid pressure is important.

At a depth h beneath the surface of a liquid of uniform density there exists a pressure *due to the weight of the liquid*. This can be expressed as

$$p = hdg$$

The absolute pressure at such a point must also take into account the pressure p_0 at the surface, such as atmospheric pressure.

$$p = hdg + p_0$$

Atmospheric Pressure. Due to the fact that the gaseous atmosphere is a fluid, atmospheric pressure exists at its base, i.e., at the earth's surface. Its normal value is approximately 14.7 lb/in.^{2} (2120 lb/ft^2 or 1,010,000 dynes/cm^2). This corresponds to the pressure at the base of a column of water 34 ft high, or a column of mercury (density 13.6 g/cm^3) 76 cm high. The barometric column (mercury barometer) is a vertical glass tube whose lower end is inserted into a cistern of mercury exposed to the atmosphere, and from whose upper end all air has been withdrawn so as to form a vacuum.

Pascal's Principle. If pressure is applied to an incompressible fluid (liquid) it is transmitted equally throughout the entire fluid at the same level. This means in the case of a liquid in communicating vessels that a force applied to a surface of the liquid where the area of cross section is small will be multiplied at some other place (at the same level as the first) where the area of cross section is large. This point is illustrated in Fig. 46.

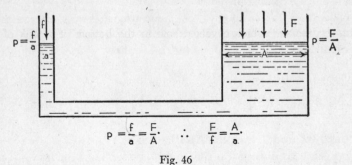

$$p = \frac{f}{a} = \frac{F}{A}. \quad \therefore \quad \frac{F}{f} = \frac{A}{a}.$$

Fig. 46

This multiplication of force by the transmissibility of pressure equally throughout a liquid at a given level, in direct proportion to the ratio of the areas, is called *Pascal's Principle*. It accounts for the action of hydraulic presses, brakes, and similar devices in which small forces overcome large forces.

Problem

A water pipe in which water pressure is 80 lb/in.^{2} springs a leak, the size of which is .01 sq in. in area. How much force is required of a patch to "stop" the leak?

Solution

This problem is quickly recognized as simply an application of the definition of fluid pressure. Commencing with the definition

$$p = \frac{f}{a}$$

it follows that: $\qquad f = pa$

Substituting values: $\quad f = 80 \times .01 = .8 \text{ lb} \quad Ans.$

Comment. Although this problem seems trivial, the method of approach is important. Always start with the defining equations of whatever concepts are involved. Note that it is not a matter of learning a formula (such as $f = pa$), but rather a matter of using a defining equation and deducing from it the information desired. Here the situation was so obvious that it was not even necessary to reduce the units to basic ones, although such a procedure is ordinarily recommended for the sake of clarity.

Problem

What is the force due to the liquid only acting on a circular plate 2 in. in diameter which covers a hole in the bottom of a tank of oil 4 ft high, if the specific gravity of oil is .5?

Solution

By definition, the pressure at the bottom of this tank is

$$p = \frac{f}{a}$$

whence $\qquad\qquad f = pa$

But the pressure at the base of a liquid is given by

$$p = hdg + p_0$$

Yet the pressure due to the liquid only is

$$p - p_0 = hdg$$

Therefore $\qquad\qquad f = hdga$

Substituting values, after reducing to basic units, and realizing also that specific gravity $S = \dfrac{d}{d_w}$, where $d_w = \dfrac{62.4}{32}$ slugs/ft^3:

$$f = (4)(.5)\left(\frac{62.4}{32}\right)(32)(\pi)\left(\frac{1}{12}\right)^2$$
$$= 2.72 \text{ lb} \quad Ans.$$

Problem

A rectangular cistern 6 ft $\times$ 8 ft is filled to a depth of 2 ft with water. On top of the water is a layer of oil 3 ft deep. The specific gravity of the oil is .6. What is the absolute pressure at the bottom, and what is the total thrust exerted on the bottom of the cistern?

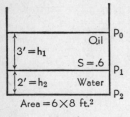

Area = 6 $\times$ 8 ft.^{2}

Fig. 47

Solution

Draw diagram and recognize the problem to deal with pressure at the base of a column of liquid—actually at the base of a column upon which there is a second column.

$$\therefore \quad p_2 = h_2 d_w g + p_1 + p_0 \qquad \text{where } p_0 \text{ is atmospheric pressure}$$
$$= h_2 d_w g + h_1 d_0 g + p_0 \qquad \text{where } d_0 \text{ is the density of oil}$$

Substitute values, where $d_w = \dfrac{62.4}{32}$ slugs/ft^3

$$d_0 = S_0 d_w = (.6)\left(\frac{62.4}{32}\right) \quad \text{and} \quad p_0 = 14.7 \text{ lb/in.}^2 = (14.7)(144) \text{ lb/ft}^2$$

$$\therefore \quad p_2 = (2)\left(\frac{62.4}{32}\right)(32) + 3(.6)\left(\frac{62.4}{32}\right)32 + 14.7(144)$$
$$= 124.8 + 112.3 + 2120 = 237.1 + 2120$$
$$= 2357 \text{ lb/ft}^2 = 16.4 \text{ lb/in.}^2 \quad Ans.$$

The total thrust on the bottom is due to the liquids and the atmosphere.

$$\therefore \quad f = p_2 A = (2357)(6 \times 8) = 113,000 \text{ lb} \quad Ans.$$

Problem

A certain hydraulic press consists of a cylinder 3 in. in diameter, and a smaller cylinder $\frac{1}{2}$ in. in diameter, filled with oil. A closely fitting piston is free to move in contact with the oil in the larger cylinder while another piston fits the smaller cylinder. In order for the larger piston to overcome a force of 12,000 lb, how much pressure must be developed in the oil? How much force must be applied to the smaller piston? If this is done by the use of a lever 15 in. long whose fulcrum is 1 in. from the piston end, what force is required at the end of the arm?

Solution

This is a straightforward application of Pascal's Principle.

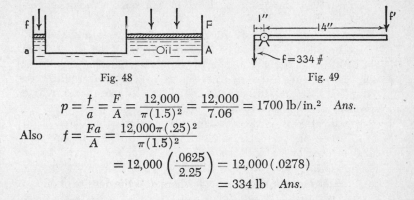

Fig. 48   Fig. 49

$$p = \frac{f}{a} = \frac{F}{A} = \frac{12,000}{\pi(1.5)^2} = \frac{12,000}{7.06} = 1700 \text{ lb/in.}^2 \quad Ans.$$

Also $\quad f = \dfrac{Fa}{A} = \dfrac{12,000\pi(.25)^2}{\pi(1.5)^2}$

$$= 12,000 \left(\frac{.0625}{2.25}\right) = 12,000(.0278)$$

$$= 334 \text{ lb} \quad Ans.$$

By the use of a lever

$$f'(14) = 334(1) \quad \therefore \quad f' = \frac{334}{14} = 23.9 \text{ lb} \quad Ans.$$

Problems: Matter and Fluid Concepts (Density, Pressure)

1. To what height would the liquid in a barometer rise if the atmospheric pressure were 10^6 dynes/cm² and the density of the liquid were 6.0 g/cc?

2. In a hydraulic press the area of the smaller piston is 3 sq in. and that of the larger piston is 100 sq in. The force exerted on the smaller piston is 100 lb. (a) What force is exerted on the larger piston? (b) What pressure is exerted on the smaller piston? (c) If the pressure were to be increased by 10 lb/sq in. on the small piston, what would be the increase in pressure on the large piston?

3. In a hydraulic press the two pistons have areas of 2 sq in. and 10 sq in. respectively. The pressure on the small piston is 40 lb/sq in.

(a) What is the pressure on the large piston? (b) What is the force acting on the large piston? (c) What is the mechanical advantage of the device?

4. Calculate the total force in pounds, exerted on the base of a cylindrical tank 40 ft high, 10 sq ft in cross section, and filled with oil of specific gravity = .8. Neglect the effects of the atmosphere.

5. What is the pressure due to glycerine alone at the bottom of a square tank filled with glycerine if the tank is 6 ft high and 3 ft on each side? Density of glycerine = 2.5 slugs/ft^3.

6. What force must the glass on an evacuated television tube withstand if it has a surface area of 626 in.^{2}? Assume standard atmospheric pressure.

7. A dam 150 ft long and 12 ft wide is inclined at an angle of 60° to the horizontal. What force is exerted on it by the water when the depth of the water is such as to just reach the top?

8. To what pressure in pounds per square inch is a diver subjected at a depth of 200 ft beneath the surface of sea water whose specific gravity is 1.03?

9. A cylindrical glass bottle of height 10 in. and sectional area of 50 sq in. has a narrow neck of 1 sq in. cross section. The bottle is filled to overflowing with water so that a rubber stopper, to be inserted in the neck, has to be forced. A tap of 5 lb on the stopper by a hammer produces how much force on the entire bottle?

10. A gash is torn in the side of a ship beneath the surface of the water by a collision. If the average depth of the gash is 20 ft and its cross-sectional area is 15 sq ft, what force of water must be overcome by a patch rigged to meet the emergency?

Archimedes' Principle of Buoyancy. When a body is immersed in a fluid, the pressure forces on the entire surface of the body do not balance out to zero because those on the lower surfaces exceed those on the upper surfaces as a result of the differences in pressure with fluid level. This difference in force constitutes a buoyant force, whose value is the weight of the displaced fluid. This is *Archimedes' Principle.* Obviously, if the specific gravity of the liquid is less than that of the body immersed in it, the body will sink to the bottom of the container so as to be completely immersed if there is sufficient liquid. On the other hand, if the specific gravity of the liquid is greater than that of the body, the latter will sink only to such a depth that the weight of the displaced liquid will equal the weight of the body; i.e., it will float only partly submerged.

Problem Procedures

When dealing with buoyancy problems it is recommended that a diagram be drawn, that the body in question be isolated, and that the vertical forces acting on it be considered. If the body is in equilibrium, all upward forces (including the buoyant force B) will balance all downward forces (including the weight W).

Recall that the weight of a body can be expressed as

$$W_b = m_b g = V_b d_b g$$

and that the buoyant force on a completely submerged body is expressible as

$$B = W_l = m_l g = V_l d_l g = V_b d_l g$$

Recall also that

$$Spg_b = \frac{d_b}{d_w} = \frac{\dfrac{m_b}{V_b}}{\dfrac{m_w}{V_w}}$$

where the subscripts b, l, and w refer to body, liquid, and water respectively. Moreover, the buoyant force on a partially submerged body is expressible as

$$B = W_l = m_l g = V_l d_l g = W_b = m_b g = V_b d_b g$$

where W_l, m_l, V_l refer to the liquid displaced by the partially submerged body.

Problem

What is the apparent loss of weight of a cube of steel 2 in. on a side submerged in water, if the specific gravity of the steel is 6.0?

Solution

Since the apparent loss of weight is merely the buoyant force, which is the weight of the water displaced, it is merely necessary to compute the weight of a cube of water 2 in. on a side.

$$W = Vdg = \left(\frac{2}{12}\right)\left(\frac{2}{12}\right)\left(\frac{2}{12}\right)\left(\frac{62.4}{32}\right)(32) = .29 \text{ lb (approx.)} \quad Ans.$$

Problem

How many cubic feet of life preserver of specific gravity .3, when worn by a boy of weight 125 lb and having a specific gravity .9, will just support him $\frac{8}{10}$ submerged in fresh water of which 1 cu ft weighs 62.4 lb?

Solution

In this problem the boy b is $\frac{8}{10}$ submerged while the life preserver p must be completely submerged to give the maximum buoyancy.

Fig. 50

The weight of the boy W_b and the weight of the preserver W_p acting downward are just balanced by the buoyant force of the preserver B_p and the buoyant force of the boy B_b.

$$B_b + B_p = W_b + W_p$$

But $\quad B_b = \frac{8}{10}V_b d_w g \qquad$ and $\qquad W_b = V_b d_b g = 125$ lb

$\qquad B_p = V_p d_w g \qquad$ and $\qquad W_p = V_p d_p g$

Also $\quad V_b = \dfrac{W_b}{gd_b} \qquad$ and $\quad d_b = S_b d_w$ and $d_w = \dfrac{62.4}{g}$

where all the symbols above are self-evident.

$$\therefore \quad \frac{8}{10}V_b d_w g + V_p d_w g = W_b + V_p d_p g$$

$$\frac{8}{10}\frac{125}{gd_b}d_w g + V_p \cancel{d_w}\cancel{g} = 125 + V_p \frac{3}{10}\cancel{d_w}\cancel{g}$$

$$V_p = \dfrac{\frac{8}{10}(125)\frac{d_w}{d_b} - 125}{\frac{3}{10}d_w g - d_w g} = \dfrac{\frac{8}{10}(125)\frac{d_b}{S_b d_y} - 125}{-\frac{7}{10}d_w g} = \dfrac{\frac{8}{10}\left(\frac{125}{\frac{9}{10}}\right) - 125}{-\frac{7}{10}\frac{62.4}{g}g}$$

$$= \dfrac{125 - \frac{8}{9}(125)}{\frac{7}{10}(62.4)} = \dfrac{\frac{1}{9}(125)}{\frac{7}{10}(62.4)} = \dfrac{13.9}{43.7} = .32 \text{ cu ft (approx)} \quad Ans.$$

Comment. Although short cuts may be introduced in this problem for the sake of a quicker answer, the forgoing solution represents a straightforward logical approach. Only the arithmetic and algebraic details seem involved, yet in reality the steps follow directly, and all possible confusion in reasoning is eliminated. The student can take comfort in the realization that he can be absolutely sure of himself at all times. There is no reason to wonder if he has correctly remembered and applied some "tricky" formula. On the other hand, he must really understand the defining equations for all the concepts involved.

Problems: Archimedes' Principle of Buoyancy

1. A loaded ship displaces 3000 tons of fresh water. What weight of sea water would it displace? (Specific gravity of sea water = 1.03, and weight of 1 cu ft of fresh water = 62.4 lb.)
2. A floating log of wood displaces 10 cu ft of fresh water. What is the weight of the log? (1 cu ft of water weighs 62.4 lb.)
3. A submerged rock displaces 10 cu ft of water. What is the buoyant force on the rock? (Weight of 1 cu ft of water is 62.4 lb.)
4. What is the volume of a 10,000 g block of silver whose specific gravity is 10.0?
5. A metal object whose weight in air is 2.0 lb is then weighed while suspended in oil of specific gravity .8. (a) If the apparent weight while in oil is 1.6 lb, what is the weight of the displaced oil? (b) If the object is weighed while suspended in water, what will be the apparent weight in water?
6. A uniform block of metal of specific gravity 5.0 floats partially submerged in mercury of specific gravity 13.6. What fraction of the block is submerged?
7. A wooden raft 6 ft × 4 ft × $\frac{1}{2}$ ft has a specific gravity of .75. How much load can it carry in fresh water to be just barely submerged?

8. Using a beam balance it is noted that 450 g are needed to balance an object in air. When this same object is suspended in water only 400 g are needed, whereas 415 g are needed if it is suspended in an unknown liquid. What is the specific gravity of this liquid?

9. What is the smallest volume of ice (specific gravity .9) that will support a 120 lb boy in fresh water?

10. A certain life belt is made of material having a specific gravity of .20. If the average specific gravity of a person is 1.05, and if it is desired to float a 150 lb man with $\frac{1}{10}$ of his volume above water, what volume must the life belt have?

Boyle's Law of Gas Pressure. As noted earlier, gases as well as liquids are fluids, and the laws of fluids apply also to them. Since they are very compressible, however, they display certain properties not common to liquids. It is an observable fact that if a given mass of gas is compressed (at constant temperature) the product of the pressure and the volume is practically constant. This is known as Boyle's Law, which stated mathematically is

$$pV = \text{constant}, \qquad \text{or} \qquad \frac{p_1}{p_2} = \frac{V_2}{V_1} \text{ at constant temperature}$$

The Ideal Gas. It has been observed further that whereas the law holds very nearly exactly for gases of simple chemical structure like hydrogen and helium, it does not hold quite so exactly for gases of more complicated chemical structure. Indeed it holds approximately for all gases but actually does not hold absolutely for any known gas. This fact has suggested the concept of the "ideal gas," a fictitious substance for which Boyle's Law would hold absolutely. It has been responsible for much of the information discovered about the nature of the structure of matter, as the relatively small discrepancies from Boyle's Law by various gases have been correlated with the increasing complexities of the substances according to their positions in the chemical table of the elements.

Problem Procedures

Here as in many other realms of Physics, problem solving involves very little more than recognizing applications of Boyle's

Law. If a problem situation stipulates that the temperature remains constant while the absolute pressure or the volume of a gas changes, then Boyle's Law is suggested. It may be, however, that the pressure, instead of being given directly in the problem, may have to be calculated from certain relationships involving the data given.

Problem

Bubbles of air escaping from a cylinder of compressed air at the bottom of a pond 34 ft deep are observed to expand as they rise to the surface. Approximately how much do they increase in volume if it can be assumed that there is no change in temperature?

Solution

Recognizing this as a situation involving Boyle's Law, it follows that

$$\frac{V_t}{V_b} = \frac{p_b}{p_t}$$

where the subscripts b and t refer to bottom and top respectively. But p_t is atmospheric pressure (1 atmosphere = 14.7 lb/in.^{2}).

Moreover, the pressure at the base of 34 ft of water is 2 atmospheres (atmospheric pressure plus an additional atmosphere due to the water).

$$p_b = 2 \text{ atmospheres } (29.4 \text{ lb/in.}^2)$$

It follows that $\dfrac{V_t}{V_b} = \dfrac{2}{1}$ *Ans.*

Problems: Boyle's Law

1. Compressed air at an absolute pressure of 100 lb/in.^{2} in a tank whose volume is 10 cu ft will expand to how many cubic feet at atmospheric pressure (15 lb/in.^{2})?
2. What volume of air at atmospheric pressure of 15 lb/in.^{2} must be compressed into a tank of capacity 50 ft^3 and at the same temperature, if the gauge pressure is to be 90 lb/in.^{2}?
3. Water is forced into an airtight storage tank until it has compressed the air in the tank to $\frac{1}{4}$ of its original volume. What is the air pressure in the tank as it would be indicated by a gauge?
4. The volume of a certain automobile tire when inflated is 4500 cu in.

If the air in the tire at gauge pressure of 30 lb/in.^{2} is allowed to escape, what volume does it amount to at atmospheric pressure?

5. A cylindrical tube closed at one end is lowered vertically with its open end down until it just touches the bottom of a lake. When the tube is withdrawn, it is found that the water had just half-filled the tube. How deep is the lake?

Fluids in Motion. Although the study of fluids at rest (Hydrostatics) is within the reach of beginning students because of the relative simplicity of the relationship between the concepts, the study of fluids in motion (Hydraulics) is much more difficult because of the complexity of the concepts. Fluid motion is usually accompanied by turbulence, which is very complex and erratic. If the study is limited, however, to the more or less ideal situation called steady flow or nonturbulent flow, a few basic concepts can be appreciated without serious difficulty.

Steady flow, sometimes called *streamline flow,* is characterized by the fact that at a given point in the fluid the velocity remains constant with time. Of course the velocity may, and usually does, differ from place to place. In the case of steady flow the *law of continuity* holds. This states that

$$a_1 v_1 = a_2 v_2$$

where a_1 and v_1 refer to the area of cross section and velocity at one place, while a_2 and v_2 refer to the area of cross section and the velocity at some other place. In a tube constricted at some point, as in Fig. 51, it follows that the fluid is speeded up where

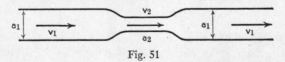

Fig. 51

the area of cross section is reduced, and then slowed down again where the original cross-sectional area is encountered.

Bernoulli's Principle. At a constriction in a tube, the increase in velocity necessarily requires an increase in the kinetic energy per unit volume of the fluid ($KE = \frac{1}{2}mv^2$). This can happen only if there is a decrease in the energy associated with the pressure, with the result that the pressure is reduced at places of constrictions, because of the principle of the conservation of energy. In the case of a horizontal tube, or pipe, and under conditions of no turbulence, Bernoulli's Principle states that

$$\tfrac{1}{2}dv_1{}^2 + p_1 = \tfrac{1}{2}dv_2{}^2 + p_2$$

A common application of this principle is found in the laboratory aspirator which serves as a vacuum pump. Water rushing through a constriction produces a reduction in pressure at a side tube. Thus Bernoulli's Principle is a consequence of the law of continuity combined with the law of the conservation of energy. See Fig. 52.

Problem

A Venturi meter is a device by means of which the velocity of flow of a fluid can be measured. The following diagram illustrates the operation of one type of such a meter. If the value of A is 10 times that of a, and colored water is used in the U-tube, what is the velocity v_1 of flow of water when the difference in levels h is 6 in.? See Fig. 53.

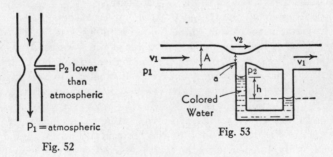

Fig. 52

Fig. 53

Solution

The operation of the above meter, according to the diagram, is obviously based upon Bernoulli's Principle. The velocity of flow at the constriction is increased so as to reduce the pressure sufficiently to cause the difference in levels of the fluid in the indicator tube. By the law of continuity

$$Av_1 = av_2 = 10av_1$$
$$v_2 = 10v_1$$

By Bernoulli's Principle

$$\tfrac{1}{2}dv_1{}^2 + p_1 = \tfrac{1}{2}dv_2{}^2 + p_2$$
$$p_1 - p_2 = \tfrac{1}{2}d(v_2{}^2 - v_1{}^2) = \tfrac{1}{2}d(100v_1{}^2 - v_1{}^2)$$
$$= \tfrac{1}{2}d(100v_1{}^2) \ \text{(approx)}$$

But $\qquad p_1 - p_2 = hdg$
$$hdg = \tfrac{1}{2}d(100)v_1{}^2$$
$$v_1{}^2 = \frac{2g}{100}\,h = .64h$$
$$v_1 = .8\sqrt{h}$$

Substituting $h = 6$ in. $= .5$ ft:

$$v_1 = .8(.7^+) = .56^+ \ \text{ft/sec} \quad Ans.$$

Problems: Fluids in Motion

1. Water flows through a horizontal pipe. The pipe is completely filled and streamline flow is assumed. At point A the speed of the water is 2 ft/sec and at point B the speed is 5 ft/sec. The cross-sectional area at A is 3 sq ft. (a) What is the cross-sectional area at B? (b) How many cu ft per second flow past point A?

2. A 40 ft *vertical* drain pipe 3 sq in. in cross-sectional area that carries water from a roof to the ground is plugged at the lower end. (a) What is the pressure due to water at the lower end when the pipe is full of water? (b) This pipe is connected to a horizontal one of 2 sq in. in cross-sectional area. If finally the water breaks through the obstruction and the velocity of the water is 50 ft/sec in the 3 sq in. pipe, what is the velocity in the 2 sq in. pipe? (Assume smooth flow.) (c) In which pipe is the pressure greater (the bottom of the 3 sq in. pipe or the horizontal 2 sq in. pipe)? (d) What is the force, due to the water, exerted on the obstruction in question before it is forced out of the pipe?

3. A 6 in. diameter water main includes a short segment which tapers down to 3 in. diameter and then flares out to 6 in. again. If the discharge rate of the system is 100 cu ft per minute and if the pressure in the main portion of the line is kept at 60 lb/in.^{2}, (a) what is the pressure in the constriction, and (b) how does the velocity of flow at the constriction compare with that in the rest of the line?

8

Wave Motion and Sound

The study of elasticity is not complete without a consideration of the propagation of deformations through deformable media. This is known as wave motion. Applications of wave phenomena are readily found in that branch of Physics known as sound. Consequently, wave concepts and acoustical phenomena are logically studied together.

Wave Motion. Various types of waves are recognized, such as *transverse waves*, *longitudinal waves*, *torsional waves*, etc., according to whether the vibrations of the medium are perpendicular to, parallel to, or rotational with respect to the direction of propagation. They are most easily represented by a diagram in which the displacement of the medium is plotted as a function of time. In the case of the simplest type of vibration, the simple harmonic type considered earlier, the graph appears as in Fig. 54, in which certain wave terminology is suggested. The *wave*

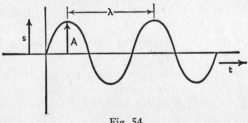

Fig. 54

length λ is the distance between two successive corresponding points, such as from crest to crest. The *amplitude* A is the maximum value of the displacement from the equilibrium position.

The *frequency* n of the wave motion is the number of complete waves that pass a given point per second. The period t of the wave motion is the reciprocal of the frequency. Thus

$$t = \frac{1}{n}$$

The basic wave relation is given by the relation

$$v = n\lambda$$

where v is the *velocity* of the wave.

Waves can be *reflected* from boundaries between different media. They also can be *refracted*, i.e., they experience a change in velocity as they pass from one medium into another medium if the elastic characteristics and the densities of the media are different. Furthermore, waves can be superimposed with the result that interference effects occur. If two similar waves are superimposed in such manner that they are exactly in phase with each other, they reinforce one another. But if the two waves are exactly out of phase with each other, they interfere destructively, i.e., they nullify one another. If two similar waves of almost, but not quite, the same frequency are superimposed, the result is the *beat* phenomenon, in which they alternately reinforce and destroy one another with a *beat frequency* exactly equal to the numerical difference in their frequencies.

An interesting interference phenomenon is observed when two waves exactly similar, except that one travels in one direction and the other in exactly the opposite direction with identical speed, are superimposed. The result is the *standing* or *stationary wave*, which displays *nodes* (regions of no disturbance) and *loops* (regions of maximum disturbance).

This phenomenon can be demonstrated by vibrating transversely a cord fixed at both ends, as in Melde's experiment. Dependent on the tension in the cord, the length of the cord, the mass per unit length of the cord, and the frequency of the vibration, is the number of nodes and loops obtained. The relation governing the phenomenon is

$$n = \frac{1}{2l} \sqrt{\frac{N^2 T}{m}}$$

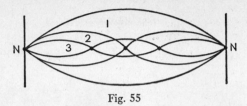

Fig. 55

where N is the number of loops, T is the tension, l is the length, m is the mass per unit length, and n is the frequency. If the cord vibrates with only a single loop or segment, it is said to be vibrating in its fundamental mode. The other modes are known as *harmonic overtones*. The fundamental mode is known as the first harmonic. When $N = 2$, the mode of vibration is said to be the second harmonic, etc.

When a source of waves approaches an observer, the frequency of the wave motion appears to increase. In general, whenever there is relative motion of source to observer along the line joining them, the observed frequency n' obeys the following relation

$$n' = n \frac{v \pm v_o}{v \mp v_s}$$

where v is the velocity of the wave, v_o is the velocity of the observer, n is the frequency of the wave, and v_s is the velocity of the source. This is *Doppler's Principle*.

Problem Procedures

As in practically all the more descriptive and less mathematical topics in Physics, the problems are generally limited to applications of the few relationships capable of mathematical formulation, such as the wave equation $v = n\lambda$, the equation dealing with standing waves in cords

$$n = \frac{1}{2l} \sqrt{\frac{N^2 T}{m}}$$

or Doppler's Principle $\qquad n' = n \dfrac{v \pm v_o}{v \mp v_s}$

Problem

What is the wave length of the sound wave emitted by a standard 440 cycles per second tuning fork?

Solution

Realizing that the relation necessary for the solution of this problem is $v = n\lambda$, and assuming the velocity of sound to be 34,000 cm/sec, or approx. 1100 ft/sec:

$$\lambda = v/n = \tfrac{1100}{440} = 2.5 \text{ ft (approx)} \quad Ans.$$

Problem

What would be the required tension in the cord in a Melde's apparatus to produce a standing wave pattern of 6 segments, by a cord 200 cm long, having a total mass of 36 g, if it is vibrated by an electric device having a frequency of 100 cycles per second?

Solution

Since the equation for the standing wave pattern in the Melde's apparatus is

$$n = \frac{1}{2l} \sqrt{\frac{N^2T}{m}}$$

this problem is solved simply by substituting the proper values in the equation, and solving for T.

$$4n^2l^2 = \frac{N^2T}{m}$$
$$T = \frac{4mn^2l^2}{N^2} \qquad m = \frac{36}{200}$$
$$\therefore \quad T = \frac{4 \cdot 36 \cdot 100^2 \cdot 200}{36}$$
$$= 800(10,000) = 8,000,000 \text{ dynes}$$

This is the weight of approx. 8.2 kg *Ans.*

Problems: Wave Motion

1. A vibrating source sends waves 27 cm in wave length along a rope with a velocity of 600 cm/sec. Find the frequency of the source.
2. An oscillator of 444 vib/sec and a tuning fork of 440 vib/sec are sounded together. What is the beat frequency?
3. The fundamental frequency of a piano string is 264 vib/sec. What is the frequency of the fourth harmonic?
4. The velocity of longitudinal waves in a certain sample of steel has

been found to be approximately 40,000 cm/sec. If the density of this steel is taken as 9 g/cm³, what is the value of Young's Modulus for this sample of steel?

5. An oscillator emitting waves with a frequency of 256 cycles/sec is moved toward an observer at a speed of 44 ft/sec. If the wave velocity is 1100 ft/sec, what is the frequency appreciated by the observer?

Sound. Wave motion is exemplified by sound. Although sound is ordinarily recognized by its auditory effects, it may be described more generally as the propagation of mechanical vibrations through elastic media.

Sources of sound include vibrating objects of all sorts, such as tuning forks, whistles, vibrating strings, diaphragms, and air columns. If the frequency of the vibration is within the range of approximately 20 cycles to 20,000 cycles the sound is more or less audible.

Sound waves travel in air, which is a fluid and as such is incapable of withstanding strains of the shear type. Therefore, sound waves in air are examples of longitudinal or compressional waves, in constrast to transverse waves. The velocity of sound in air depends upon the temperature as follows:

$$v = v_o \sqrt{1 + \tfrac{1}{273}t}$$

where v_o is the velocity under standard conditions (0° C and 76 cm Hg), and t is the temperature in Centigrade degrees. The velocity of sound is readily measured by setting up a standing wave pattern, as will be described presently, whereby the wave length can be measured, and since the frequency is known, we have $v = n\lambda$.

Standing Sound Waves in Pipes. A pipe open at one end and closed at the other end is referred to as a *closed* pipe. One open at both ends is called an *open* pipe. A source of sound held at the open end of a closed pipe of length l will produce standing waves in the pipe if its frequency and the length of the pipe are such that a node can be formed at the closed end while a loop is formed at the open end. The longest standing wave (fundamental) that can thus be formed has a wave length

$$\lambda_c = 4l \text{ (closed pipe)}$$

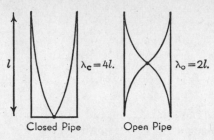

Fig. 56

For an open pipe, however, there must be a loop at each open end. This means that there must be at least one node between the two loops, so that the longest fundamental wave length has to be

$$\lambda_o = 2l$$

Harmonic overtones can also be set up in these pipes providing the above conditions are satisfied, namely, that nodes occur at the closed end, and loops are formed at open ends. Because of symmetry, all the harmonics can be formed commencing with the fundamental (first harmonic) in open pipes, each end being the same. In the case of closed pipes, however, only the odd-numbered harmonics can be formed. Thus the first overtone in the case of a closed pipe is the third harmonic.

Problem

A sound is returned as an echo from a distant cliff in 4 sec. How far away is the cliff, assuming the velocity of sound in air to be 1100 ft/sec?

Solution

Although this type of problem is frequently encountered, it is actually a simple rate problem and does not involve special acoustical concepts.

The sound travels twice the distance s.

$$2s = v \times t$$
$$s = \frac{v \times t}{2} = \frac{1100 \times 4}{2} = 2200 \text{ ft} \quad Ans.$$

Characteristics of Sound. The three outstanding characteristics of sound are *loudness, pitch,* and *quality.*

Loudness depends upon the intensity of the sound wave, which in turn varies with the second power of the amplitude. The ratio of the loudness of two sounds is proportional to the ratio of the logarithms of their intensities. Loudness levels are expressed in *decibels,* ranging from 10–20 decibels for a quiet room, to 100–120 near a jet airplane motor.

Pitch is the auditory effect of frequency. For a normal person, sounds are audible in a frequency range 16–20,000 cycles per second near the upper levels of loudness.

Quality depends upon the presence of overtones. Sources like tuning forks which emit so-called pure notes devoid of overtones are conspicuous for their lack of quality. In the preceding section it was pointed out that open pipes (like organ pipes) can emit all harmonic overtones whereas closed pipes emit only those whose frequencies are odd multiples of the fundamental. Therefore open organ pipes have more quality than closed organ pipes.

Problem

When two tuning forks are sounded simultaneously, a beat note of 5 cycles per second is heard. If one of the forks has a known frequency of 256 cycles per second, and if a small piece of adhesive tape fastened to this fork reduces the beat note to 3 cycles per second, what is the frequency of the other fork?

Solution

Although this problem involves sound, it is nothing more than an illustration of the phenomenon of beats. When two similar waves are superimposed, the beat frequency represents the numerical difference in their frequencies.

$$\therefore \quad n = 256 \pm 5$$

where n represents the unknown frequency.

It appears that n has two possible values, either 251 or 261. If, however, the beat frequency is reduced to 3 cycles per second when the frequency of the standard fork is reduced by loading it, the unknown frequency must be less rather than more than 256.

$$\therefore \quad n = 251 \quad Ans.$$

Problem

As a train whose whistle has a fundamental frequency of 260 cycles per second approaches a station, the pitch of the whistle heard by an observer on the platform is increased by 22 cycles. After the train passes the platform the pitch is observed to drop to a frequency 22 cycles below normal. What is the speed of the train?

Solution

This problem illustrates the Doppler effect for a stationary observer and an approaching source.

$$n' = n \, \frac{v + v_o}{v - v_s}$$

where v is the velocity of the sound wave, v_o is the velocity of the observer, v_s is the velocity of the source, n is the frequency of the source, and n' is the frequency appreciated by the observer.

Assume $v = 1100$ ft/sec for sound. The problem states that

$$n' - n = 22 \qquad \text{and} \qquad v_o = 0; \; n = 260$$

$$n' = \frac{nv}{v - v_s} \qquad \therefore \quad n'v - n'v_s = nv$$

$$n'v_s = (n' - n)v$$

$$v_s = \frac{n' - n}{n'} \, v = \frac{22}{282} \, (1100) = 86 \text{ ft/sec} \quad Ans.$$

Problem

What will be the frequencies of the first and the second overtones of a closed pipe of length 2 ft? What will be the frequencies of the first and the second overtones of an open pipe 2.5 ft long? Will there be any common beat frequency between these overtones?

Solution

The first and second overtones of a closed pipe will be the third and the fifth harmonic respectively, i.e., frequencies 3 times and 5 times that of the fundamental.

For the closed pipe the fundamental frequency is $n_f = v/\lambda = v/4l$, where l is the length of the closed pipe.

$$\therefore \quad n_f = \frac{1100}{4 \cdot 2} = \frac{1100}{8} = 137.5$$

$$n_{1st} = 3(137.5) = 412.5 \atop n_{2nd} = 5(137.5) = 687.5 \Bigr\} \ Ans.$$

In the case of the open pipe, the first and the second overtones will be the second and third harmonics, i.e., frequencies 2 times and 3 times that of the fundamental.

For the open pipe: $n_f' = \dfrac{v}{\lambda} = \dfrac{v}{2l} = \dfrac{1100}{2 \cdot 2.5} = 220$

$$n'_{1st} = 2(220) = 440 \atop n'_{2nd} = 3(220) = 660 \Bigr\} \ Ans.$$

The frequency difference for the first overtones is $440 - 412.5 = 27.5$. The frequency difference for the second overtones is $687.5 - 660 = 27.5$.

Thus a common beat frequency is 27.5 cycles per sec. *Ans.*

Problems: Sound

1. Two sounds are emitted at the same time. The beat note has a frequency of 4 vib/sec. One of the sounds has a frequency of 150 vib/sec. What is a possible value for the other?

2. In an organ pipe of length l closed at one end and open at the other: (a) What is the wave length of the fundamental? (b) What is the wave length of the first overtone?

3. In an organ pipe of length l and open at both ends: (a) What is the wave length of the fundamental? (b) What is the wave length of the first overtone?

4. An organ pipe closed at one end has a length of 40 cm. The velocity of sound in air is 34,400 cm/sec. (a) What is the wave length of the fundamental note emitted by this pipe? (b) What is the frequency of the fundamental note? (c) What is the wave length of the second overtone emitted?

5. What is the wave length of the fundamental associated with a closed pipe 90 cm long?

6. The first overtone of a closed pipe has the same frequency as the third overtone of an open pipe. If the open pipe has a length of 3 ft, what is the length of the closed pipe? Assume velocity of sound = 1100 ft/sec.

7. A sound is followed by an echo 4 sec later. How far away is the reflecting surface?

8. What is the frequency of the third overtone of a closed pipe 50 cm long if the velocity of sound is taken as 34,000 cm/sec?

9. If a train passes a station at 80 mph where a signal gong emits a note whose frequency is 400 cycles per second, (a) what pitch does a passenger on the train observe as the train approaches the gong? (b) as the train recedes from the gong?
10. Two open pipes, one 2 ft long and the other 2.1 ft long, sound together. What is the frequency of the beat note that is heard?

Part III.

Heat

9

Nature of Heat and Temperature

Although heat was formerly thought to be a material substance (calorific fluid), it has for something more than a hundred years been considered as a form of energy. Temperature is, qualitatively speaking, a property which governs the direction of flow of heat. In a sense, it is also a measure of the intensity or degree of heat. Heat can flow of its own accord from regions of high temperature to regions of low temperature. Thus heat and temperature are not to be confused, the measurement of heat and the measurement of temperature being entirely different procedures. The latter utilizes certain thermal properties of matter, whereas an understanding of what heat is requires an appreciation of the structure of matter in order to visualize heat as a form of energy.

Kinetic Theory of Matter. The present-day concept of matter pictures it as made of molecules, atoms, electrons, protons, and neutrons as well as a number of less important and recently discovered unstable and short-lived particles including mesons, pions, strange particles, etc. The atom is the smallest unit of a chemical element; the molecule is the smallest unit of a chemical substance or compound. Molecules are in a state of random motion. In the gaseous state they have considerably more freedom than in the liquid state, and more in the liquid than in the solid state.

The randomness of the motion suggests the motion of bees in a swarm, having individually no apparent regularity but having velocities capable of statistical treatment. Since each moving molecule has mass, it has kinetic energy due to its motion. The *total energy* thus associated with the *random motion* of the molecules of matter is what is meant by *heat* being a form of energy. In other words, the addition of heat to a body increases the thermal agitation of its molecules.

On the basis of the kinetic theory, the following relationship can be shown to hold for the pressure of a gas resulting from the bombardment of the walls of the container by molecules:

$$p = \tfrac{1}{3}nmv^2$$

where n is the number of molecules per unit volume, m is the mass of a molecule, and v is the average molecular translational velocity. This theory can also justify Boyle's Law and Avogadro's number (Chemistry), and explains Brownian motion.

Temperature is defined as the average kinetic energy of translation associated with the random motion of the molecules.

Comment. The preceding material on the nature of heat does not lend itself readily to illustration by problems at the elementary level. It is included primarily to enable the student to form mental pictures of some of the concepts which follow in the study of heat, in order that the problems will not seem like mere substitution in formulas.

Thermometry. There are many thermal properties of matter, but simple linear expansion is utilized in the mercury-in-glass thermometer, in which the mercury expands from a glass bulb into a connected glass capillary tube. Two temperature scales are in common use, the Centigrade and the Fahrenheit. They are both related to the freezing point and the boiling point of water. On the Centigrade scale the freezing temperature of water is arbitrarily designated as 0°, and the boiling temperature at normal atmospheric pressure is designated as 100°, thus providing 100 equally spaced units of temperature between these limits. On the Fahrenheit scale these points are designated as +32° and +212° respectively, with the intervening range thus subdivided into 180 equal units. This means that 180 Fahrenheit degrees corresponds to 100 Centigrade degrees.

Problem Procedures

Problems in thermometry often reduce to exercises in converting temperatures from one scale to another. These should not be thought of as requiring formulas for the solutions, but should be solved on the basis of a knowledge of the nature of the temperature scales. Realizing that each Fahrenheit degree is $\frac{5}{9}$ (same as $\frac{100}{180}$) the size of each Centigrade degree, and that 0° C corresponds to +32° F, conversion problems should be handled somewhat as follows.

Problem

What temperature on the Centigrade scale corresponds to the common room temperature of 68° F?

Solution

Because $68 - 32 = 36$, note that +68° F is 36 Fahrenheit degrees above the freezing point of water. 36° F above the freezing point corresponds to $\frac{5}{9}$ (36) $= 20°$ C above the freezing point. But since the freezing point is 0° C, this temperature is +20° C. Thus +68° F $= 20°$ C. *Ans.*

Problem

What Fahrenheit temperature corresponds to −40° Centigrade?

Solution

−40° C is 40 Centigrade degrees below the freezing point of water. Now 40° C $= \frac{9}{5}(40) = 72°$ F. But 72° F below the freezing point, which is 32° F, is 40° F below 0° F $(72 - 32 = 40)$. Thus −40° C $=$ −40° F. *Ans.*

Comment. This is the only temperature reading which is common to the Centigrade and the Fahrenheit scales.

Thermal Expansion of Solids and Liquids. It is found experimentally that when most objects in the form of long thin rods are heated, they expand in length in such a manner that, over a more or less limited range of temperatures, the fractional increase in length per degree of temperature is fairly constant. This constant is known as the *thermal coefficient of linear expansion.* Mathematically this is stated: $l = l_0(1 + \alpha t)$ or $l - l_0 = l_0 \alpha t$ where l_0 is the length at 0° C, t is the Centigrade temperature, and α is the coefficient of linear expansion. Values of α with respect to 0° C can be found in tables of physical constants for all common substances. It is to be noted that since α is a fractional change in length per unit of temperature, its numerical value depends upon the temperature scale used. For example, in the case of steel:

$$\alpha = .000012/°C \qquad \text{or} \qquad \tfrac{5}{9}(.000012) = .0000067/°F$$

Problem

In the design of a modern steel bridge, provisions must obviously be made for expansion. How much does this amount to in the case of a bridge two miles long which is subjected to temperatures ranging from $-40°$ F to $+110°$ F, assuming an average expansion coefficient of .000012/°C?

Solution

Starting with the defining equation $\alpha = l - l_0/l_0 t$ or $l - l_0 = \alpha l_0 t$, it is clear that either the temperature range must be expressed in Centigrade degrees, or the coefficient must be converted to Fahrenheit degrees. The temperature range $-40°$ F to $+110°$ F is $150°$ F or $\tfrac{5}{9}(150) = 83.3°$ C.

$$\therefore \quad l - l_0 = (.000012)(2 \times 5280)(83.3) = 1.2 \times 10^{-5} \times 2 \times 5.280$$
$$\times 10^3 \times 8.33 \times 10 = 105.6 \times 10^{-1} = 10.56 \text{ ft} \quad Ans.$$

Volume Expansion. Expansion in three dimensions is characterized by the relation

$$V = V_0(1 + \beta t)$$

where V is the volume at temperature t, V_0 is the volume at $0°$ C, and β is the thermal coefficient of volume expansion. For practical purposes

$$\beta = 3\alpha$$

Gases. In the case of gases both volume and pressure change with temperature. If the pressure is kept constant

$$V = V_0(1 + \beta t)$$

where β has the value $.003667 = \dfrac{1}{273}$ (approx)

If the volume is kept constant

$$p = p_0(1 + bt)$$

where p is the pressure at temperature t, p_0 is the pressure at $0°$ C, and b is the pressure-temperature coefficient. For gases b has the same value as β, namely .003667 (approx). This suggests that at the temperature $-273°$ C, the volume and the pressure of a gas vanish, and that $-273°$ C is absolute zero.

The Absolute Temperature Scale. The above considerations suggest an absolute scale of temperatures on which the freezing point of water is $+273°$ A and the boiling point of water is $+373°$ A. (A refers to absolute temperature and is commonly represented by the capital T. Absolute temperature is also often designated as $°K$ in honor of Lord Kelvin.)

Thus
$$T_C = t_C + 273$$
$$T_F = t_F + 460$$

where the subscripts refer to Centigrade and Fahrenheit respectively.

Although absolute zero has never been reached in the laboratory, temperatures within a few thousandths of a degree absolute have been attained. Temperatures lower than absolute zero are inconceivable.

General Gas Law. Combining the volume-temperature relation, the pressure-temperature relation, and Boyle's Law of pressure–volume, the general gas law is obtained. It takes the form

$$\frac{p_1 V_1}{T_1} = \frac{p_2 V_2}{T_2}$$

or
$$\frac{pV}{T} = R, \text{ the so-called gas constant}$$

This general relationship is called the equation of state for the ideal gas. In other words, the state in which matter exists, i.e., whether gaseous, liquid, or solid, depends upon its p–V–T values.

Problem

A 20 gallon automobile gasoline tank is filled exactly to the top at 0° F just before the automobile is parked in a garage where the temperature is maintained at 70° F. How much gasoline is lost due to expansion as the car warms up? Assume the coefficient of volume expansion of gasoline to be .0012/ °C.

Solution

From the defining equation, it follows that $V - V_0 = V_0 \beta t$. Furthermore the temperature range 0°–70° F corresponds to $\frac{5}{9}(70) = \frac{350}{9} = 39°$ C.

$$\therefore \quad V - V_0 = 20(.0012)(39) = 2 \times 1.2 \times 3.9 \times 10^{-1} = .94 \text{ gal}$$

Ans.

Problem

How deep is a pond if bubbles forming at the bottom quadruple their size in rising to the top, and if the temperature at the bottom is assumed to be 10° C and the temperature at the surface is observed to be 20° C?

Solution

Recognize this problem as involving the general gas law, where the pressure at the top is atmospheric pressure p_0 and the pressure at the bottom exceeds atmospheric by an amount corresponding to the given depth of water. All temperatures must be expressed on the absolute scale.

$$\frac{p_0 V_0}{T_0} = \frac{pV}{T}$$

$p_0 = 1$ atmosphere $\qquad \frac{V_0}{V} = 4 \qquad T_0 = 20 + 273 \qquad T = 10 + 273$

Solving for p $\qquad p = p_0 \frac{V_0 T}{V T_0} = 1(4)\left(\frac{283}{293}\right) = 3.87$ atmospheres

$3.87 - 1 = 2.87$ atmospheres due to water. But $p = hdg$ for a liquid of density d and depth h.

For water, it is recalled that one atmosphere corresponds to approximately 34 ft, and so it is unnecessary to substitute values in the above equation. Therefore 2.87 atmospheres corresponds to 2.87(34) = 97.6 ft (approx).

$$\therefore \quad h = 97.6 \text{ ft (approx)} \quad Ans.$$

Problems: Thermometry, Expansion, General Gas Law

1. When the reading on a Fahrenheit thermometer is 102° F, what is the reading on a Centigrade thermometer?
2. An air conditioner lowers the temperature of a room 15° F. What is the change in degrees Centigrade?
3. The gas in a cylinder pushes back a piston and does 1440 ft-lb of work. This is done at a constant pressure of 2880 lb/ft². What is the change in volume?
4. How much does a 30 ft section of railroad rail expand when its temperature changes from −20° F to 100° F? The coefficient of linear expansion for steel is $7 \times 10^{-6}/°F$.
5. A certain mass of gas at 300° K with a volume of 10^6 cu cm and an absolute pressure of 10^6 dynes/sq cm is compressed and heated so that the temperature becomes 400° K and the volume 10^4 cu cm. What is the final absolute pressure?
6. What temperature on the Fahrenheit scale corresponds to 31° C, the critical temperature of carbon dioxide?
7. How much allowance must be made for the expansion of a steel bridge 1 mile long which experiences a range in temperatures between +110° F and −40° F if the coefficient of linear expansion is $7 \times 10^{-6}/° F$?
8. A spherical ball of steel has a volume of 400 cu cm at 0° C. By how much does the volume increase as the temperature is raised to 100° C if the coefficient of linear expansion is $11 \times 10^{-6}/°C$?
9. The air pressure in an automobile tire is checked at −10° C and found to be 26 lb/in.². After the car is run and the temperature rises to +30° C, what is the pressure if the volume of the casing is assumed to remain constant?
10. A balloon containing 1000 cc of gas at 40° C and 5 lb/in.² gauge pressure is cooled to 20° C. If the gas is allowed to contract to a volume of 950 cc, what will its pressure be?

10

Calorimetry

It is convenient to consider heat as something that can be measured in terms of the rise in temperature which it can produce. The amount of heat required to raise the temperature of one gram of water one degree Centigrade is designated as *one calorie*. The amount of heat required to raise the temperature of one pound of water one degree Fahrenheit is referred to as one *British Thermal Unit* (Btu). These values are averaged over the temperature range from the freezing point to the boiling point.

$$1 \text{ Btu} = 252 \text{ calories}$$

In terms of energy one calorie represents 4.18 joules (1 joule = 10^7 ergs), and one Btu represents 778 ft-lb. Recall that one joule equals 10^7 ergs (dyne-cm), or one newton-meter.

From the definition of the calorie (or Btu), it is clear that heat quantities are measured with reference to water. The amount of heat required to raise the temperature of one gram of any substance one degree Centigrade is known as the *heat capacity* (c) of that substance. The ratio of the heat capacity of any substance to the heat capacity of water (unity) is called the *specific heat* (s) of that substance. Thus specific heat is a dimensionless quantity.

If a given amount of heat raises the temperature of m grams of a substance from t_0 to t, then a numerical measure of this amount is

$$H = ms(t - t_0) \text{ calories}$$

Determination of Specific Heat by Method of Mixtures.

When a heated object of known mass and temperature is dropped into a known mass of a liquid, such as water, the rise in temperature of the liquid is an indication of the specific heat of the object.

Assuming that the heat lost by the object as it cools equals the heat gained by the water and its container:

$$m_l s_l(t_h - t_f) = m_w(t_f - t_i) + m_c s_c(t_f - t_i)$$

where m_l is the mass of the heated object, s_l is its specific heat, t_h is its original temperature, t_f is its final temperature (the same as the final temperature of the water), m_w is the mass of the water, t_i is the initial temperature of water and container, and m_c and s_c are the mass and specific heat respectively of the container. By substituting observed values for all these quantities except s_l, the value of s_l can be computed. Of course the method is applicable to other liquids than water if the specific heat of the liquid used is known.

Problem Procedures

In problems involving heat quantities it is frequently profitable to equate the heat lost by bodies in the process of cooling to the heat gained by bodies in the same system that are warming up. The amount of heat in each case is given by multiplying the mass of each body by the specific heat of the substance by the change in temperature, using Centigrade degrees when mass is expressed in grams, or Fahrenheit degrees when mass is expressed in pounds. This gives the amount of heat in calories or Btus respectively. Values of specific heat may be found in tables of physical constants, unless the problem is that of determining the value of the specific heat of some body in terms of the data given in the problem.

Problem

A 100 g block of copper ($s_{cu} = .095$) is heated to 95° C and is then plunged quickly into 1000 g of water at 20° C in a copper container whose mass is 700 g. It is stirred with a copper paddle of mass 50 g until the temperature of the water rises to a steady final value. What is the final temperature?

Solution

The heat lost by the hot copper block as it cools to temperature t_f is

$$m_{block}s_{cu}(t_{95} - t_f) = 100(.095)(95 - t_f)$$

The heat gained by the water, the container, and the paddle is

$$(m_{water} + m_{container}s_{cu} + m_{paddle}s_{cu})(t_f - t_{20})$$
$$[1000 + 700(.095) + 50(.095)](t_f - 20)$$

Equating the heat lost to the heat gained:

$$100(.095)(95 - t_f) = [1000 + 700(.095) + 50(.095)](t_f - 20)$$

Regrouping and solving for t_f:

$$(9.5)(95 - t_f) = (1000 + 66.5 + 4.75)(t_f - 20)$$
$$902.5 - 9.5t_f = 1071.3(t_f - 20) = 1071.3t_f - 21430$$
$$22330 = 1081t_f$$
$$t_f = \frac{22330}{1081} = 20.6° \text{ C} \qquad Ans.$$

Problem

500 g of alcohol at 75° C are poured into 500 g of water at 30° C in a 300 g glass container ($s_{glass} = .14$). The mixture displays a temperature of 46° C. What is the specific heat of alcohol?

Solution

Express the heat lost by the alcohol.

$$m_{alcohol}s_{alcohol}(t_{75} - t_f)$$
$$500s(75 - 46) = 500s(29) = 14500s$$

Express the heat gained by the water and the glass container.

$$m_{water}(t_f - t_{original}) + m_{glass}s_{glass}(t_f - t_{original})$$
$$500(46 - 30) + 300(0.14)(46 - 30)$$
$$500(16) + 42(16) = 542(16) = 8670$$

Equating the heat lost to the heat gained:

$$14500s = 8670$$
$$s = \tfrac{8670}{14500} = .598 \qquad Ans.$$

Change of State. Within the range 0° C–100° C the addition of heat to water raises the temperature at the rate of one degree Centigrade per gram per calorie, under normal circumstances. At 100° C, the normal boiling temperature, a break occurs in the temperature-heat curve; this break is associated with the change of state from liquid water to steam. To complete the conversion 540 calories per gram are required while the temperature remains at 100° C. Thus to boil water at 100° C it is not sufficient merely to raise the temperature to 100° C. An additional amount of heat must be added. This is called the *heat of vaporization* (540 calories per gram for water).

Similarly, at 0° C an amount of heat equal to 80 cal/gram must be withdrawn from water to freeze it after the temperature has been reduced to the freezing point. This is called the *heat of fusion* (80 cal/gram for water).

This also means that to melt ice 80 cal/gram must be added at 0° C to produce the conversion, and to condense steam 540 cal/gram are released in the conversion process.

Problem

A 200 g ice cube is placed in 500 g of water at 20° C. Neglecting the effects of the container, what is the resultant situation?

Solution

Note that a cube of ice at 0° C will lower the temperature of the water, and that for every 80 calories absorbed by the ice, one gram will be melted without any change in temperature. If the heat given off by the 500 g of water cooling to 0° C exceeds the amount necessary to melt the 200 g of ice, then the water will not cool to 0° C. If, however, it is less than sufficient to melt all 200 g of ice, only a fraction of the ice will be melted, and the resultant temperature will be 0° C.

The amount of heat that must be withdrawn from the water to lower its temperature to 0° C is

$$500 \times 20 = 10000 \text{ cal}$$

The amount of ice that 10000 calories will melt at 0° C is

$$\frac{10000}{80} = 125 \text{ g}$$

This is less than 200 g, the original amount of ice. Therefore a 75 g block of ice finds itself floating in water at 0° C.   *Ans.*

Problem

500 g of lead shot at a temperature of 100° C are poured into a hole in a large block of ice. If the specific heat of lead is .03, how much ice is melted?

Solution

As 500 g of lead cool from 100° C to 0° C, $500(.03)(100) = 1500$ calories are released. This will melt:

$$\frac{1500}{80} = 18.75 \text{ g of ice}$$

Since presumably all the ice present is not melted, the temperature of the water will not rise above 0° C, i.e., the ice and the water will remain in equilibrium.

18.75 g melted   *Ans.*

Problem

Steam at atmospheric pressure is passed into a pail of ice water in which a 5 lb block of ice is floating in 10 quarts of ice water. How many pounds of steam are necessary to melt the ice and raise the temperature of the water to 70° F, no steam to be wasted in the process?

Solution

Since the data are given in English units (assuming one pint of water to have a mass of one pound) it will be desirable to express the heats of vaporization and fusion of steam, water, and ice in Btu per pound, and to express temperatures in Fahrenheit degrees.

$$540 \text{ cal/gram} = \frac{540}{252}(454) = 973 \text{ Btu/lb}$$
$$80 \text{ cal/gram} = \frac{80}{252}(454) = 144 \text{ Btu/lb}$$
$$10 \text{ qt water} = 20 \text{ lb}$$

Temperature range from steam point to ice point $= 180° \text{ F}$

Equate heat released by condensation of m lb of steam, added to the heat lost in m lb of condensed steam cooling to 70° F, on the one hand, to the heat required to melt 5 lb of ice and to raise 20 lb plus 5 lb of water from 32° F to 70° F.

$$m(973) + m(212 - 70) = (5)144 + (20 + 5)(70 - 32)$$
$$m(1115) = 1670$$
$$m = \tfrac{1670}{1115} = 1.5 \text{ lb} \qquad Ans.$$

Problems: Calorimetry and Change of State

1. An iron ball having a mass of 320 g was heated in a furnace and then dropped into 300 g of water in a copper vessel of 110 g mass at 20° C. The final temperature was 80° C. The specific heat of iron is .105, and that of copper .092. (a) How much heat was absorbed by the water? (b) How much heat was absorbed by the copper vessel? (c) What was the temperature of the furnace?

2. How much water could be cooled from 20° C to 5° C, if 10 g of ice were placed in a glass of water, and if no heat exchange with the surroundings took place? Neglect the heat capacity of the glass.

3. A silver spoon of mass 320 g was heated in a furnace and then dropped into 300 g of water in a copper vessel of 110 g mass at 20° C. The final temperature was 50° C. The specific heat of silver is .055, and that of copper .092. (a) How much heat was absorbed by the water? (b) How much heat was absorbed by the copper vessel? (c) What was the temperature of the furnace?

4. 800 calories of heat are added to a block of copper whose mass is 1000 g and whose specific heat is .09. How many degrees is its temperature raised?

5. A block of ice having a mass of 30 g is dropped into 1000 g of water at 100° C. If the heat capacity of the container is neglected, what is the final temperature of the mixture?

6. A copper object having a mass of 600 g is heated to 99° C and then dropped into 500 g of water in a 100 g copper calorimeter at 19° C. The temperature of the water rises to what value if the specific heat of copper is taken to be .092?

7. If the temperature of steam is 100° C, how many grams will be required to melt 30 kg of ice whose temperature is 0° C?

8. If 30 g of ice at 0° C are dropped into 300 g of water at 25° C in a copper calorimeter of mass 100 g, what is the temperature of the water after the ice has melted?

9. How much heat is required to boil away 250 cu cm of water at 100° C?

10. It is proposed to raise the temperature of 1 cu m of water in an iron tank from 10° C to 30° C by the introduction of steam at 100° C. Taking the specific heat of iron to be .09, how many grams of steam will be required? Assume the mass of the tank to be 10 kg.

11

Heat Transfer and Thermodynamics

There are three modes by which heat may be transferred from one place to another. These modes are called conduction, convection, and radiation.

Conduction. Conduction is a process in which thermal agitation of molecules is passed along throughout a substance. Thus heat is literally propagated through a physical medium. The rate of heat transfer by conduction is given by the expression

$$\frac{H}{t} = \frac{kA(t_2 - t_1)}{X}$$

where t represents time, k is a proportionality constant (coefficient of thermal conductivity) which depends upon the particular substance in question, A is the cross section area, t_2 and t_1 are the temperatures at either end, and X is the length of the conductor. When H is expressed in calories, A is expressed in cm^2, X is expressed in cm, and temperatures are expressed on the Centigrade scale. If, on the other hand, H is expressed in Btus, A is in ft^2, X is in ft, and temperatures are expressed on the Fahrenheit scale. Values of k are to be found in tables of physical constants.

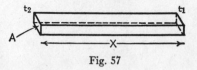

Fig. 57

Problem Procedure

Most problems involving heat conduction at the level of the first-year Physics course seem to be simple applications of the conduction formula stated above. Care must be taken to express the data in units consistent with those for which k is given.

Problem

On either side of a pane of window glass, temperatures are 70° F and 0° F. How fast is heat conducted through such a pane of area 2500 cm² if the thickness is 2 mm?

Solution

From tables it is found that $k = .0015$ in cgs units for glass. This suggests that the temperature difference be expressed in Centigrade degrees.

$$70° \text{ F} = \tfrac{5}{9} (70) = 38.9° \text{ C}$$
$$\frac{H}{t} = \frac{kA(t_2 - t_1)}{X}$$

where $A = 2500 \text{ cm}^2$ $\quad t_2 - t_1 = 38.9° \text{ C} \quad X = .2 \text{ cm}$

$$\therefore \quad \frac{H}{t} = \frac{.0015(2500)38.9}{.2} = 729 \text{ cal/sec} \quad Ans.$$

Problem

A refrigerator has 40 sq ft of surface walls that are 2 in. thick. The inside temperature is maintained at 40° F while the outside temperature is 80° F. If the walls are constructed of wood, at what rate does heat leak through them? What would the rate be if the refrigerator walls were made of solid glass?

Solution

This problem suggests the heat conductivity equation $H/t = kA(t_2 - t_1)/X$, but the data are not given in cgs units. From tables, however, values of k for wood and glass are found to be as follows:

$$k_{\text{wood}} = 9 - 3 \quad (\text{Btu-in./hr ft}^2 \text{ F }°)$$
$$k_{\text{glass}} = 6 \quad (\text{Btu-in./hr ft}^2 \text{ F }°)$$

Since the average value of k_{wood} is the same as that of k_{glass}, namely 6 Btu-in./hr ft² F °, the answer to the second part of the question is obvious. The rate is the same in each case.

$$\frac{H}{t} = \frac{6(40)(40)}{2} = 4800 \text{ Btu/hr} \quad Ans.$$

Comment. This is enough heat to melt approx. 33 lb of ice per hour, since 80 cal/g = 144 Btu/lb.

Convection. Convection is the mode of heat transfer in which heat is carried from one place to another by a physical agent such as a flowing fluid (liquid or gas). Usually convection currents are set up, but because of the complexity of fluid flow, except for the ideal situation encountered in streamline flow, numerical problems in convection are too complex for the first-year Physics course.

Radiation. Radiation is a wavelike mode of heat transfer in which heat is treated as an electromagnetic-wave propagation through the luminiferous ether (same as radio and light, but of different wave-length range). The rate of radiation from an ideal so-called "black body" surface into free space is expressible as follows

$$W = SAT^4$$

where W is expressed in joules/sec (same as watts), A is the surface area expressed in cm^2, and T is absolute Centigrade temperature (4.18 joules are equivalent to 1 calorie). If the radiating surface is at temperature T_1 and the surroundings are at temperature T_2, the net rate of radiation is given by the equation

$$W = SA(T_1{}^4 - T_2{}^4)$$

S is the so-called "black body" constant and has the value 5.7×10^{-12} joules per second, per square centimeter, per degree absolute to the fourth power.

Radiation is explained by the now famous Quantum Theory, which postulates that radiant energy is always emitted in quanta of various sizes, but never as fractional parts thereof. Thus radiation is considered to be noncontinuous and intermittent, in contrast to the concepts of wave propagation. Hence arose the so-called conflict between the wave and the corpuscular theories of radiation that has been resolved only in modern times by the concepts of wave mechanics, which are beyond the scope of this book.

Problem

How much faster does a cup of coffee cool one degree from 100° C than from 30° C in a room at 20° C?

Solution

Although at first glance it might seem that the rate of cooling is the same because of the fact that the same amount of heat is involved (1° change in temperature in each case), the fact is that the rate of cooling is faster at the higher temperature because the rate depends upon the *fourth* power of the absolute temperature.

$$W = SA(T_1{}^4 - T_2{}^4)$$

Although the coffee is not an ideal "black body," its radiation constant can be treated as proportional to that of the ideal "black body."

$$S_1 = \sigma S$$
$$\therefore \quad W = \sigma SA(T_1{}^4 - T_2{}^4)$$

For the hot coffee $\quad W_1 = \sigma SA(373^4 - 293^4)$

where $\quad T_1 = 100 + 273 \quad$ and $\quad T_2 = 20 + 273$

For the cold coffee $\quad W_2 = \sigma SA(303^4 - 293^4)$

$$\therefore \quad \frac{W_1}{W_2} = \frac{373^4 - 293^4}{303^4 - 293^4} = \frac{(193 - 73.6) \times 10^{16}}{(84.3 - 73.6) \times 10^{16}} = \frac{119.4}{10.7} = 11.2$$

Therefore the hot coffee cools 1° C at the rate of 11.2 times faster than the cold coffee. *Ans.*

Thermodynamics. Two basic laws of thermodynamics are usually considered in first-year Physics.

(1) The first law of thermodynamics is the law of the conservation of energy applied to heat. It states that 4.18 joules of mechanical energy can be converted into 1 calorie of heat. (It does not state that the opposite is true.) Therefore, in the development of heat by the expenditure of energy:

$$4.18 \text{ joules} \leftrightharpoons 1 \text{ cal}$$

4.18 joules per calorie is known as the mechanical equivalent of heat and is often represented by the letter J (Joule).

$$J = 4.18 \text{ joules/cal}$$

(2) The second law of thermodynamics deals with the irreversibility of natural processes. There are several different ways of stating the law. One statement is that heat of its own accord flows only from regions of high temperature to regions of low temperature.

One conclusion which is drawn from the second law of thermodynamics is that the efficiency of an ideal engine in converting heat into mechanical energy is limited by the working temperature range of the engine. Furthermore, since no real engine can be more efficient than (actually never as efficient as) an ideal engine, the efficiency of all heat engines is thus limited.

$$\text{Eff.} = \frac{T_1 - T_2}{T_1}$$

where T_1 is the absolute temperature of the boiler (intake), and T_2 is the absolute temperature of the exhaust. Only if the temperature of the exhaust were absolute zero would the ideal engine be 100% efficient.

Problem

If a 50 g bullet traveling at 40,000 cm/sec is stopped by a steel plate, how much heat is generated, on the assumption that the energy is completely converted into heat?

Solution

Recognize this situation as one in which the kinetic energy ($\frac{1}{2}mv^2$) of the bullet is converted into heat at the rate of 4.18 joules per calorie.

$$\therefore \quad \tfrac{1}{2}mv^2 = JH$$
$$\tfrac{1}{2}(50)(40,000)^2 = 4.18H$$
$$H = \frac{25(1,600,000,000)}{4.18} = 96 \times 10^8 \text{ ergs} = 960 \text{ joules} \quad Ans.$$

Problem

Gas expanding in a gas engine moves a piston in a cylinder and does 2000 ft-lb of work at a constant pressure of 3000 lb/sq ft. How much does the volume change, and how much heat could be developed in the process?

Solution

This problem obviously deals with the equivalence of energy and heat. Whereas work is defined as

$$W = fs$$

it can also be expressed for an expanding gas as $W = pV$, where W is the amount of work done, as a gas at constant pressure expands an amount V.

(*Note!* $p = f/a$ and $pV = (f/a)V = fs$ dimensionally.)

$$\therefore \quad pV = JH = W$$

$$\therefore \quad V = \frac{W}{p} = \frac{2000}{3000} = .667 \text{ cu ft}$$

Also $\qquad H = \frac{2000}{778} \qquad (4.18 \text{ joules/cal} = 778 \text{ ft-lb/Btu})$

$$= 2.57 \text{ Btu} \quad Ans.$$

Problem

What is the maximum efficiency of a steam engine if the temperature of the input steam is 175° C and the temperature of the exhaust is 75° C?

Solution

Problems like this are solved by direct substitution in the formula for the efficiency of an ideal engine.

$$\text{Eff.} = \frac{T_1 - T_2}{T_1} = \frac{(175 + 273) - (75 + 273)}{175 + 273}$$

$$= \frac{448 - 348}{448} = \frac{100}{448} = 22.3\% \quad Ans.$$

Problems: Heat Transfer and Thermodynamics

1. The bottom of a copper boiler is .1 cm thick and has a cross-sectional area of 400 cm². If the upper surface is maintained at 80° C and gas flames keep the under surface at 100° C, how much heat will flow through in 5 min? The thermal conductivity of copper is .092 cal/cm-sec-°C.

2. What is the rate of heat conduction in calories per second through a plate .30 cm thick and 10^5 sq cm in area if the coefficient of thermal conductivity is .0030 cal/cm-sec-°C and if the temperature difference of the two sides is 50° C?

3. A jet engine operates between a high temperature of 650° C and an exhaust temperature of 40° C. Find its theoretical efficiency.

4. A steam engine takes heat into its cylinders from the steam boilers at a temperature of 200° C and exhausts it at a temperature of 100° C. Calculate the maximum efficiency of the engine.

5. An iron bolt 20 cm long with cross-sectional area of 5 cm² extends through the wall of a building. The temperature of the inner end is 15° C and that of the outer end is −20° C. How much heat per minute is conducted from the inside to the outside? The thermal conductivity of iron is .16 cal/cm-sec-°C.

6. A slab of copper 20 cm × 10 cm 2 cm thick is so mounted that the temperature of one face is maintained at the temperature of boiling water while the other is in contact with flowing water maintained at a temperature of 60° C. If the thermal conductivity of this material is .092 cal/cm-sec-°C, how much heat passes through the slab in 2 hours?

7. A refrigerator is insulated by being surrounded with cork board which has a thermal conductivity of approximately 9 Btu-inches per square foot per day per °F. If the surface area of the refrigerator is 32 sq ft, and the thickness of the insulation is 2 in., how much heat leaks through in one week? Assume the inside temperature to be 50° F and the outside temperature to be 80° F.

8. If the "black body" thermal radiation constant is accepted as 5.7 × 10^{-8} watt/meter Ab. Temp.[4], at approximately what rate is heat radiated from a tungsten sphere 2 mm in radius maintained at a temperature of 600° C in a relatively large vacuum chamber?

9. A given carnot engine has an efficiency of 40%. What must be the temperature of its high temperature reservoir if its exhaust temperature is 50° C?

10. A 1 hp heat engine works between the temperatures of 150° C and 50° C. If its actual efficiency is one quarter of its theoretical efficiency, how many foot-pounds of energy are absorbed per second?

Part IV.

Electricity and Magnetism

12

Electrostatics

In spite of its practical importance in the modern world, electricity is an abstract concept incapable of definition in terms of anything simpler. Electric charge is therefore a basic concept to be added to those of length, mass, and time discussed in mechanics. Charge is made evident by the attractive or repulsive force observed to exist between charged bodies. When a rod of hard rubber or ebonite is rubbed with fur, or when a rod of glass is rubbed with silk, each of these objects becomes capable of attracting to itself small bits of paper, and is said to be electrically charged. Moreover, each is observed to repel similarly treated like objects and to attract similarly treated opposite objects, i.e., charged hard rubber repels charged hard rubber but attracts charged glass. The charge on the *glass* is called *positive*, and the charge on the *hard rubber* is called *negative*.

Coulomb's Law. Electric charges obey a law discovered by Coulomb, expressible in two parts:

(a) Like charges repel and unlike charges attract.

(b) The force of attraction or repulsion between two charged bodies (assumed pointlike) depends directly upon the product of the charges and inversely upon the second power of the separation. It also depends upon the medium in which the charged bodies are located, i.e., the medium between them.

$$F = \frac{qq'}{Kr^2}$$

where q and q' represent the magnitudes of the charges, r is the separation, and K is the *dielectric constant* of the medium.

It is not uncommon to write $F = k\,(qq^1/r^2)$, where the proportionality constant k, rather than $1/K$, is used. When electrostatic cgs units are used, $1/K$ is the more convenient because K has the numerical value of the dielectric constant of the medium. The value of k in cgs units is 1 dyne-centimeter^2 per stat-coulomb^2. In mks units, however, k has the numerical value of 9×10^9 newton-meter^2 per coulomb^2, because 1 newton-meter^2 per coulomb^2 equals 9×10^9 dyne-centimeter^2 per stat-coulomb^2.

A quantity $\varepsilon_0 = 1/4\pi k = 8.85 \times 10^{-12}$ coulombs^2/newton-meter^2 is introduced to represent for vacuum a concept which corresponds to ε, a characteristic of a real substance called permittivity, such that the dimensionless ratio $\varepsilon/\varepsilon_0$ is really what is meant by the dielectric constant K of that substance, i.e., $K = \varepsilon/\varepsilon_0 = 1/k$.

Note! The quantity ε differs from K by the factor 4π, which is arbitrarily introduced to avoid this factor in the several electrical formulas which can be derived from Coulomb's Law. For vacuum $K = 1$ and $\varepsilon = \varepsilon_0$.

Electrical Units. Coulomb's Law provides a logical means of standardizing a unit of electric charge. Such a unit is that amount of charge which will repel an exactly equal amount of charge of like sign with a unit of force when the two are placed unit distance apart in vacuum (both charges considered to be pointlike). In the cgs electrostatic system this quantity of charge is called the *stat-coulomb*. The coulomb is 3×10^9 stat-coulombs. Moreover, the electronic charge is -4.80×10^{-10} stat-coulombs.

Obviously the stat-coulomb is a supposititious unit since it cannot be realized experimentally. Yet it is significant to realize that an abstract concept like electric charge can be treated quantitatively by its use. Also, called the *unit positive charge*, this concept became the basis for all quantitative considerations in electricity until the discovery of the electron, a so-called natural or experimentally realizable unit.

In mks units, it proves to be more convenient to standardize the unit of current (rate of flow of charge). Consequently the coulomb, which is the unit of charge in this system, is a derived unit, being derived from the unit of current, the ampere, which is defined in terms of one of its measurable effects. (See later discussion of electric currents and magnetism.) Thus the proportionality factor k in Coulomb's Law is not unity for vacuum when these units are used, but 9×10^9 newton-meter^2/ coulomb^2, whereas in cgs units $1/K$ has dimensions of dyne-cm^2/stat-coulomb^2 and a numerical value of unity for vacuum.

Comment. Since problem solving is the main theme of this text, the preceding discussion of electrical units is far more than academic. The use of mks units is more widespread in electricity than in mechanics where the English units have seemed more natural than either metric system. In electricity, however, there is practically no call for English units and the competition between cgs and mks units is real. The former appeal to those who emphasize the logical and historical developments of the concepts, while the latter appeal to those who emphasize the practical and the measurement side of the subject and do not mind introducing an arbitrary constant such as $k = 1/4\pi\varepsilon_0$ where $\varepsilon_0 = 8.85 \times 10^{-12}$ coulombs^2/newton-meter^2 into the relationships between the concepts (formulas). Thus

$$F = \frac{1}{K} \frac{qq'}{r^2} = k \frac{qq'}{r^2} = \frac{1}{4\pi\varepsilon_0} \frac{qq'}{r^2}$$

(cgs) $K =$ unity for vacuum   (mks) $k = 9 \times 10^9$ for vacuum   (mks) ε_0 refers to vacuum but is not unity

Electrical Nature of Matter. Since matter is now thought to be electrical in nature, it should be noted that positive and negative charge merely refers to deficiency and surplus, respectively, of electron content. Charges are not created by rubbing glass with silk, etc., but are thereby separated as the normal distribution of electrons is altered. Electrically neutral objects are to be thought of as objects with a sufficient free electron content to balance the positive charges of the nuclei of the atoms of which they are constituted. In terms of the electrostatic unit, the charge on the electron has been found to be:

1 electron $= 4.80 \times 10^{-10}$ stat-coulombs $= 1.6 \times 10^{-19}$ coulombs

Electric Field of Force. The region around a charge and in which the influence of the charge can be experienced is called the *electric* field of the charge. The intensity E of the electric field at any point is defined as the force per unit positive charge at that point. It is measured by the force that would act on the suppositious unit positive charge if it could be placed there without itself influencing the field.

$$E = \frac{F}{q}$$

Field intensity E is a vector quantity. It is expressed in dynes per stat-coulomb or newtons per coulomb depending on the system used.

Lines of Force. To aid in visualizing the distribution of field intensity, the concept of line of force is often used. It is a line which at every point indicates, by its direction, the direction the suppositious unit positive charge would move if placed there, i.e., directly away from a positive charge (Coulomb repulsion) and directly toward a negative charge (Coulomb attraction). By adopting the convention of drawing one line per square centi-

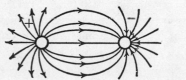 

Fig. 58

meter for each dyne per stat-coulomb of field intensity, the field intensity at a place is expressible in lines per square centimeter. Thus the line of force concept has quantitative as well as qualitative significance; both the direction and the concentration of lines are important. See Fig. 58.

In the vicinity of an isolated charge q, the field intensity can be expressed

$$E = \frac{F}{q'} = \frac{qq'}{Kr^2 q'} = \frac{q}{Kr^2} = \frac{kq}{r^2} \quad \text{where} \quad k = \frac{1}{4\pi\varepsilon_0}$$

$$= \frac{1}{4\pi \times 8.85 \times 10^{-12}} = 9 \times 10^9 \text{ mks units}$$

where q' represents the test charge. If a conductor is hollow, the field intensity everywhere inside of it is zero.

Problem Procedures

To determine the value of the electric field intensity at some point, attempt to visualize a unit positive charge placed there and then analyze the force or forces acting on it due to neighboring charge or charges, recalling that forces add vectorially. Always draw a diagram of the problem situation.

Problem

What is the intensity of the electric field 10 cm from a negative point charge of 500 stat-coulombs in air?

Solution

$E = F/q'$ where $q' = +1$ esu

By Coulomb's Law: $F = \dfrac{qq'}{Kr^2}$

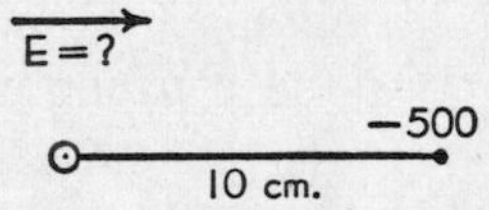

$\therefore \ E = q/Kr^2$

For air, K is approximately unity.

$\therefore \ E = \dfrac{500}{(1)(10^2)} = 5$ dynes/stat-coulomb or 5 lines/cm² pointing directly *toward* the negative charge   *Ans.*

Fig. 59

Problem

(a) Determine the field intensity midway between two identical positive charges of 200 stat-coulombs each, in oil of dielectric constant 5, if the charges are 6 cm apart.

(b) What is the answer if one charge is positive and the other is negative?

Solution

(a) Before attempting to calculate the numerical value of the field intensity E at point P due to each charge, note that if unit positive charge is placed at P, the forces acting on it because of each charge will point in opposite directions and will balance each other to give a resultant field intensity of zero.

$\therefore \ E$ at P is zero   *Ans.*

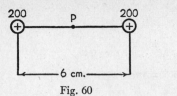

Fig. 60

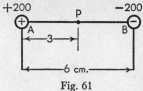

Fig. 61

(b) Unit positive charge placed at P will experience a force due to A:

$$E_1 = \frac{q}{Kr^2} = \frac{200}{5(9)} = 4.4 \text{ dynes/stat-coulomb to the right}$$

Also, due to B:

$$E_2 = \frac{q}{Kr^2} = 4.4 \text{ dynes/stat-coulomb also to the right}$$

$$\therefore \quad E = E_1 + E_2 = 8.8 \text{ dynes/stat-coulomb from} + \text{to} - \qquad Ans.$$

Problem

What must be the charge on each of a pair of pith balls suspended in air from the same point by strings 5 cm long, if they repel each other to a separation of 4 cm? Assume each pith ball to have a mass of .1 g.

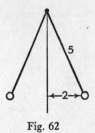

Fig. 62

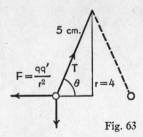

Fig. 63

Solution

Draw a vector diagram of the forces acting on one pith ball, noting that the pith ball is in mechanical equilibrium.

Consider the x-components. $\Sigma F_x = 0$.

$$\frac{-q^2}{r^2} + T \cos \theta = 0.$$

Consider the y-components. $\Sigma F_y = 0$.

$$T \sin \theta - mg = 0$$

Solve algebraically:

$$\left. \begin{array}{l} T \sin \theta = mg \\[2mm] T \cos \theta = \dfrac{q^2}{r^2} \end{array} \right\} \quad \therefore \quad \tan \theta = \dfrac{mgr^2}{q^2}$$

$$\text{and} \quad q^2 = \dfrac{mgr^2}{\tan \theta}$$

Now $\cos \theta = \frac{2}{5} = .4.$ $\qquad \therefore \qquad \tan \theta = 2.25$ (tables)

$$q^2 = \frac{(0.1)(980)(4)^2}{2.25} = \frac{98(16)}{2.25} = 697$$
$$q = \sqrt{697} = 26.4 \text{ stat-coulombs} \qquad Ans.$$

Solution in Mks Units

$$q^2 = \frac{mgr^2}{k \tan \theta}$$

where $m = 10^{-4} \text{ kg}$ $\quad r = 4 \times 10^{-2} \, m$ $\quad k = 9 \times 10^9$
$\tan \theta = 2.25$ $\quad g = 9.8 \text{ m/sec}^2$

$$\therefore \quad q^2 = \frac{10^{-4} \times 9.8 \times 16 \times 10^{-4}}{9 \times 10^9 \times 2.25} = 77.5 \times 10^{-18} \text{ coulombs}^2$$

$$\therefore \quad q = 8.8 \times 10^{-9} \text{ coulombs} \qquad Ans.$$

But 1 coulomb $= 3 \times 10^9$ stat-coulombs.

$$\therefore \quad q = 8.8 \times 10^{-9} \times 3 \times 10^9 = 26.4 \text{ stat-coulombs} \qquad Check.$$

Problems: Electric Field Intensity

1. Two isolated pointlike charges of $+400$ stat-coulombs and $+100$ stat-coulombs are 20 cm apart in air. (a) What is the magnitude of the force between the two charges? (b) What is the magnitude of the electric field intensity at a point midway between the two charges? (c) How far from the $+400$ stat-coulombs charge on a line connecting the two charges is the field intensity equal to zero? (d) If a charge of $+4$ stat-coulombs were placed at the point found

in part (c), what is the magnitude of the force that would act on it? (e) If a charge of +4 stat-coulombs were placed at the point midway between the two original charges, would it move toward the +400 stat-coulombs charge or toward the +100 stat-coulombs charge?

2. Two equal charges experience a force of repulsion of 9 dynes when they are .20 cm apart in a vacuum. What is the magnitude of the charges?

3. A pointlike charge of +12 stat-coulombs is placed at a point in space where the electric field intensity is 4 dynes/stat-coulomb. What force acts on the charge?

4. An isolated pointlike charge of +400 stat-coulombs is in air. (a) What is the magnitude of the electric field intensity at a point 10 cm from the charge? (b) If a charge of −2 stat-coulombs were placed at the point found in (a), what is the magnitude of the force it would experience? (c) In what direction would the −2 stat-coulombs charge tend to move? (d) How far from the +400 stat-coulombs charge would the field intensity be one-fourth as great as it is at the point given in (a)?

5. Two small pith balls having a mass of 1 g each are suspended from the same point by independent threads 50 cm long. They are given identical electric charge, as a result of which they repel each other to a separation of 10 cm. Calculate the charge on each pith ball.

6. Charges of +100 stat-coulombs are placed at the eight corners of a cube of 2 cm edge. What is the field intensity at the center?

7. Charges of +600 and −600 stat-coulombs are 20 cm apart in air. What force (magnitude and direction) will be experienced by a +10 stat-coulomb charge placed midway between them?

8. A repulsive force of 900 dynes is found to act on a +10 stat-coulomb charge placed 30 cm from a charge Q in an oil bath of dielectric constant 5. What is the magnitude of the charge Q?

9. Two point charges, one of -6×10^{-8} coulombs and the other of $+8 \times 10^{-8}$ coulombs are .02 meters apart in air. What is the force between them, and is it a force of attraction or repulsion?

Electrical Potential. To move an electrical charge in an electrical field from an initial point to a point of different field intensity involves work (see the definition of work in the section on mechanics).

The work required to move a supposititious unit positive charge from point A to point B is referred to as the *difference in electrical potential* between A and B. In Fig. 64 consider points

A and B at distances r_1 and r_2 respectively from a point charge of value $+q$.

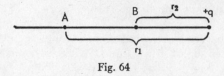

Fig. 64

A unit positive charge placed at A will experience a force

$$F_A = \frac{q \times 1}{Kr_1{}^2}$$

to the left (repulsive), whereas at B it will experience a force

$$F_B = \frac{q \times 1}{Kr_2{}^2}$$

also to the left. If it is displaced from A to B, i.e., by an amount $(r_1 - r_2)$, it will have to overcome an average force approximately equal to q/Kr_1r_2 to the left, and work must be done on it. The work per unit charge is expressed V_{BA} to indicate that at B the potential is higher than at A.

$$V_{BA} = \frac{q}{K}\left(\frac{1}{r_2} - \frac{1}{r_1}\right) = V_B - V_A$$

Absolute Potential. The above relationship states that the potential at B exceeds that at A by a certain amount. If r_1 (the distance from A to the charge q) were to become infinitely large, V_{BA} would become q/Kr_2. Thus the work required to bring a unit positive charge from infinitely far away (from outside the influence of any charges in question) to the point B is called the *absolute potential* at B. Similarly the work required to bring a unit positive charge from infinity to point A is the *absolute potential* at A. Thus the work required to bring a unit positive charge from A to B is the increase in potential of B over A.

Potential is expressed in ergs per stat-coulomb, or stat-volts. Related to the more familiar unit, the volt (not yet defined),

$$1 \text{ stat-volt} = 300 \text{ volts}$$

Potential V is a scalar quantity, unlike the vector field intensity E. Thus the absolute potential at a point due to a number of neighboring point charges is the algebraic sum of the absolute potentials due to each separately. Thus

$$V = \Sigma \frac{q}{Kr}$$

where, if q is expressed in stat-coulombs, r in centimeters, K is unity for vacuum (or air), V is expressed in stat-volts.

Or
$$V = \Sigma \frac{kq}{r}$$

where, if $k = 9 \times 10^9$, q is expressed in coulombs, r in meters, V in volts.

Note! This checks with the facts that 1 stat-volt equals 3×10^2 volts, 1 centimeter equals 10^{-2} meters, $k = 9 \times 10^9$, and 1 coulomb equals 3×10^9 stat-coulombs.

Note also that in this text all formulas involving K (dielectric constant) suggest the use of cgs units, whereas mks units are suggested when the formulas include

$$k = \frac{1}{K} = 9 \times 10^9 \text{ (numerically)}$$

As a conductor is charged, its potential changes. It becomes higher as positive charge is added, and lower as negative charge is added. A conducting surface is an equipotential surface, because charges move freely in conductors. The potential V at the surface of a charged spherical conductor of radius r is the same as if all the charges were concentrated at the center point, i.e., $V = Q/Kr$, where Q represents the total charge, and K is the dielectric constant of the medium in which the spherical conductor is located. Everywhere inside a spherical conductor the potential is the same as at the surface.

Problem

(a) What is the value of the potential midway between two identical positive charges of 200 stat-coulombs each, in oil of dielectric constant 5, if the charges are 6 cm apart?

(b) What is the answer if one charge is positive and the other is negative?

Solution

(a) Draw a diagram.
Both charges contribute to the value of the potential at point P. Thus

$$V_P = \Sigma \frac{q}{Kr} = \frac{+200}{5(3)} + \frac{200}{5(3)} = 26.7 \text{ stat-volts} \quad Ans.$$

(b) If one of the charges is negative

$$V_P = \frac{+200}{5(3)} - \frac{200}{5(3)} = 0 \quad Ans.$$

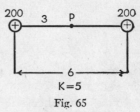

Fig. 65

Comment. Note that whereas the field intensity E is zero midway between identical charges (second problem of preceding section) the potential V is not, and vice versa for the midpoint between opposite equal charges.

Problem

What is the difference in potential between the points A and B in Fig. 66? Which is at the higher potential, and how much work will be required to move 20 stat-coulombs from the one to the other?

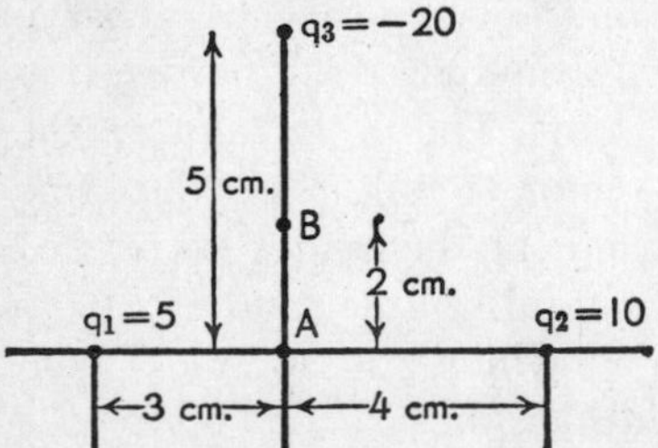

$$\begin{cases} q_1 = +5 \text{ stat-coulombs} \\ q_2 = +10 \text{ stat-coulombs} \\ q_3 = -20 \text{ stat-coulombs} \end{cases}$$

Fig. 66

Solution

The absolute potential at A is

$$V_A = +\tfrac{5}{3} + \tfrac{10}{4} - \tfrac{20}{5}$$
$$= +1.67 + 2.50 - 4.00$$
$$= +.17 \text{ stat-volts}$$

The absolute potential at B is

$$V_B = \frac{+5}{\sqrt{3^2 + 2^2}} + \frac{10}{\sqrt{4^2 + 2^2}} - \frac{20}{3}$$

$$= \frac{5}{3.6} + \frac{10}{4.5} - \frac{20}{3}$$

$$= 1.39 + 2.22 - 6.67 = -3.06 \text{ stat-volts}$$

The difference in potential between A and B is 3.23 stat-volts   *Ans.*
V_A is higher than V_B *Ans.*
To move 20 stat-coulombs from point B to point A requires $20(3.23)$
$= 64.6$ ergs.   *Ans.*

Solution Using Mks Units

$$q_1 = \frac{5}{3 \times 10^9} = 1.67 \times 10^{-9} \text{ coulomb} \qquad q_2 = \frac{10}{3 \times 10^9}$$

$$= 3.33 \times 10^{-9} \text{ C} \qquad q_3 = \frac{-20}{3 \times 10^9} = -6.67 \times 10^{-9} \text{ C}$$

At Point A, $r_1 = 3 \times 10^{-2} m$ $\qquad r_2 = 4 \times 10^{-2} m$ $\quad r_3 = 5 \times 10^{-2} m$
At Point B, $r_1 = 3.6 \times 10^{-2} m$, $r_2 = 4.5 \times 10^{-2} m$, $r_3 = 3 \times 10^{-2} m$,
and $k = 9 \times 10^9$ (numerically)

$$\therefore \quad V_A = \Sigma \frac{kq}{r} = 9 \times 10^9 \left(\frac{1.67 \times 10^{-9}}{3 \times 10^{-2}} + \frac{3.33 \times 10^{-9}}{4 \times 10^{-2}} \right.$$
$$\left. - \frac{6.67 \times 10^{-9}}{5 \times 10^{-2}} \right)$$

$$= 9.0(55.7 + 83.2 - 133.4) = 49.5 \text{ volts}$$

$$\text{or } \frac{49.5}{300} = +.17 \text{ stat-volts} \quad \text{Check.}$$

$$V_B = 9 \times 10^9 \left(\frac{1.67 \times 10^{-9}}{3.6 \times 10^{-2}} + \frac{3.33 \times 10^{-9}}{4.5 \times 10^{-2}} - \frac{6.67 \times 10^{-9}}{3 \times 10^{-2}} \right)$$

$$= 9 \,(46.0 + 74.0 - 222) = -918 \text{ volts}$$

$$\text{or } - \frac{918}{300} = -3.06 \text{ stat-volts} \quad \text{Check.}$$

$$\therefore \quad V_A - V_B = 50.5 + 918 = 968 \text{ volts} = 3.23 \text{ stat-volts} \quad \text{Check.}$$
$$\text{Work} = 6.67 \times 10^{-9} \times 968 = 64.7 \times 10^{-7} \text{ joules} = 64.7 \text{ ergs}$$
$$\text{Check.}$$

Electrical Capacitance. The amount by which the potential of an isolated conductor is increased by the addition of a given amount of charge depends upon the geometry of the conductor. The ratio of charge to potential is referred to as the *electrical capacitance* of the conductor. Capacitance is also frequently called simply *capacity*.

$$C = \frac{Q}{V}$$

Two conductors charged by equal amounts, but by opposite sign, and separated by an insulator, have a combined capacitance greater than either charged conductor alone. The combination is called a *condenser* or a *capacitor*.

For an isolated conducting sphere of radius r in a medium of dielectric constant K

$$C = Kr = \frac{r}{k} \quad \text{where } k = \frac{1}{K} = \frac{1}{4\pi\varepsilon_0} = 9 \times 10^9 \text{ mks units}$$

For a plane parallel plate condenser consisting of 2 plane parallel plates of area A separated by a dielectric of thickness d and dielectric constant K

$$C = \frac{KA}{4\pi d} = \frac{A}{4\pi k d}$$

Formulas for the capacitances of other types of condensers are to be found in engineering tables.

Series and Parallel Combinations of Condensers. When condensers are connected in parallel (see Fig. 67) the capaci-

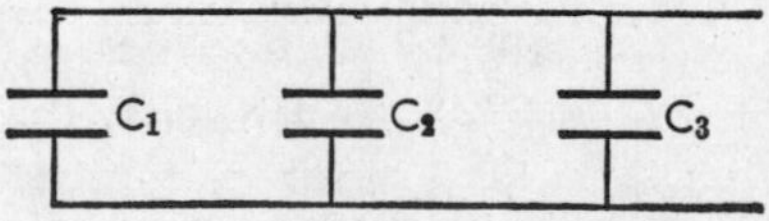

Fig. 67

tance C of the combination is given in terms of the individual capacitance as follows:

$$C = C_1 + C_2 + C_3 + \cdots \text{ (Parallel)}$$

When condensers are connected in series (see Fig. 68) the capacitance of the combination is:

$$\frac{1}{C} = \frac{1}{C_1} + \frac{1}{C_2} + \frac{1}{C_3} + \cdots \text{(Series)}$$

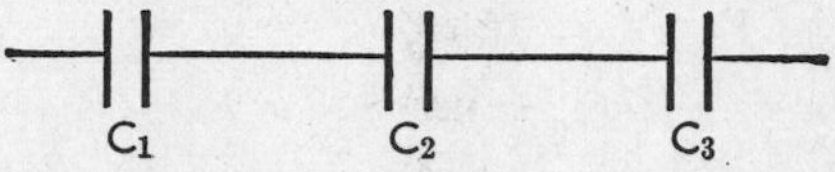

Fig. 68 C_1 C_2 C_3

Units of Capacitance. Capacitance is expressed in stat-coulombs per stat-volt, or stat-farads in cgs units, or coulombs per volt, or farads in mks units.

$$9 \times 10^{11} \text{ stat-farads} = 1 \text{ farad}$$

Condensers used in radio have capacitances of the order of micro-farads and even micro-microfarads.

Energy Stored in a Condenser. Obviously work is required to charge a condenser, since a repulsive force has to be overcome as the charge on a body is increased by additional like charge. Thus energy is stored by a charged condenser. Its amount is expressible in several ways due to the basic relationship between charge, potential, and capacitance ($C = Q/V$). Denoting energy by W

$$W = \frac{1}{2} QV = \frac{1}{2} CV^2 = \frac{1}{2} \frac{Q^2}{C}$$

When a condenser is discharged, part of this energy is sometimes released in the form of an electric spark.

Problem Procedures

Problem situations are best resolved by considering the basic nature of the concepts of *charge, potential,* and *capacitance* which are of course related by the equation

$$C = \frac{Q}{V}$$

If a condenser is involved, it should be understood that its capacitance is governed by some formula which may be assumed to be available, such as that for the common plane parallel plate condenser ($C = KA/4\pi d$) or the isolated sphere ($C = Kr$). It is generally assumed that potential at a point of known distance from point charges can be readily determined ($V = \Sigma q/Kr$), and that the formulas for condenser combinations ($C = C_1 + C_2 + C_3 + \cdots$ Parallel) and $\left(\dfrac{1}{C} = \dfrac{1}{C_1} + \dfrac{1}{C_2} + \dfrac{1}{C_3} + \cdots \text{Series}\right)$ are immediately available.

Although it cannot be denied that so-called formulas may be involved in the solutions of problems in electrostatics, it is proper to reiterate that the procedure should not be one of simply memorizing formulas. The analysis should consist of reasoning out situations from basic definitions and relationships which, of course, are often most concisely expressed mathematically. If, as in mechanics, the student is sufficiently familiar with the meanings of the relationship, he will find little difficulty in bringing them to bear on problem situations. He most certainly should not expect to find some worked-out formula listed somewhere for the complete solution of every conceivable problem situation he might encounter, but should be prepared to work with the basic relationships to serve his immediate purpose.

Problem

An isolated sphere 10 cm in radius is charged in air to 500 statvolts. How much charge is required? If this charge is then shared with another isolated sphere of 5 cm radius by connecting them together quickly with a fine wire, what is the final charge on each and what is the final potential of each?

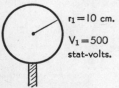

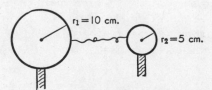

Fig. 69   Fig. 70

Solution

Part 1. Draw a diagram.

$$r_1 = 10 \text{ cm}$$
$$V_1 = 500 \text{ stat-volts}$$

Since $C = Q/V$ and since $C = Kr$ for a sphere $(K = \text{approx } 1 \text{ for air})$:

$$Q_1 = C_1 V_1 = K r_1 V_1 = 10 \times 500 = 5000 \text{ stat-coulombs} \qquad Ans.$$

Part 2.

$$r_1 = 10 \text{ cm}$$
$$r_2 = 5 \text{ cm}$$

In contact, the potential of the two spheres is the same.

$$Q = 5000 \qquad V_2 = \frac{Q}{C} \qquad \text{where} \qquad C = C_1 + C_2$$

$$\therefore \quad V_2 = \frac{5000}{10 + 5} = \frac{5000}{15} = 333.3 \text{ stat-volts}$$

whereupon the charge on each becomes

$$Q_1' = C_1 V_1 = 10(333.3) = 3333 \text{ stat-coulombs} \quad Ans.$$
$$Q_2' = 5(333.3) = 1667 \text{ stat-coulombs} \quad Ans.$$

Problem

A plane parallel plate condenser, consisting of two metal circular plates 5 cm in radius separated 1 mm in air, is charged to 300 stat-volts, whereupon it is connected in parallel to another similarly charged condenser (positive terminals connected together and negative terminals connected). How much energy would be released if the combination were discharged by a short circuit?

Solution

Diagram.
For a plane parallel plate condenser:

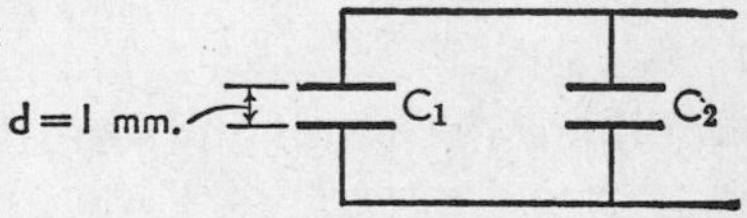

$$C = \frac{KA}{4\pi d}$$

$d = 1$ mm.

Fig. 71

$$\therefore \quad C_1 = \frac{Kr^2}{4\pi d} = \frac{5^2}{4(.1)} = \frac{25}{.4} = 62.5 \text{ stat-farads}$$

Recalling that the energy stored in a condenser is

$$W = \frac{1}{2}QV = \frac{1}{2}CV^2 = \frac{1}{2}\frac{Q^2}{C}$$

and choosing the second form because C and V are known

$$W_1 = \frac{1}{2}C_1V_1{}^2 = \frac{1}{2}(62.5)(300)^2 = 31.25(90,000) = 2,820,000 \text{ ergs}$$

But the total energy is $W_1 + W_2 = 2W_1$

$$\therefore \quad W = 2(2,820,000) = 5,640,000 \text{ ergs}$$
$$= .564 \text{ joules} \qquad Ans.$$

Solution in Mks Units

Data:

$$\text{plate radius} = 5 \times 10^{-2}\ m \quad d = 10^{-3}\ m \quad V = 9 \times 10^4 \text{ volts}$$

$$C = \frac{1}{4\pi k}\frac{A}{d} = \frac{\cancel{\pi}\, 25 \times 10^{-4}}{4\cancel{\pi}\, 9 \times 10^9 \times 10^{-3}} = 6.94 \times 10^{11} \text{ farads}$$

But 1 farad = 9×10^{11} stat-farads

$$\therefore \quad C = 6.94 \times 10^{-11} \times 9 \times 10^{11} = 62.5 \text{ stat-farads} \quad Check.$$

$$W = \frac{1}{2}C_1V_1{}^2 = \frac{1}{2}6.94 \times 10^{-11} \times 81 \times 10^8 = 28.1 \times 10^{-2} \text{ joules}$$

But 1 joule = 10^7 ergs.

$$\therefore \quad W = 28.1 \times 10^{-2} \times 10^7 = 2.81 \times 10^6 \text{ ergs}$$

$$\text{Total} \quad W = 2 \times 28.1 \times 10^{-2} = .562 \text{ joules}$$
$$= 5.62 \times 10^6 \text{ ergs} \qquad Check.$$

Problems: Potential and Capacitance

1. A pointlike charge of $+100$ stat-coulombs is in a vacuum. Point A is 10 cm to the right of the charge and point B is 20 cm to the left of the charge. What is the difference in potential between points A and B?

2. An isolated hollow metal sphere is 10 cm in radius. What is its capacitance in stat-farads?

3. A single isolated charge of +1200 stat-coulombs is in a vacuum. Points A and B are also in a vacuum, A being 10 cm from the charge and B, 15 cm from the charge. (a) What is the absolute potential at point A? (b) What is the difference in potential between points A and B? (c) How much work would be required to move a +4 stat-coulomb charge from point B to point A? (d) If the +4 stat-coulomb charge moved from point A to point B, what would be its change in potential energy?

4. An isolated sphere of conducting material is 8 cm in radius and is in air. It is charged until its total is 64 stat-coulombs. (a) What is the capacitance of the sphere? (b) What is the absolute potential of the sphere? (c) How much energy was required to charge the sphere to a total of 64 stat-coulombs?

5. Two isolated pointlike charges of +1200 stat-coulombs and −400 stat-coulombs are 20 cm apart in air. (a) What is the absolute potential at a point P midway between the charges? (b) What is the difference in potential between point P and another point Q on the line joining the charges and 5 cm from the −400 stat-coulombs charge? (c) How much work would be required to move a charge of +2 stat-coulombs from point Q to point P? (d) If the +2 stat-coulomb charge moved from point P to point Q, what would be its change in potential energy? (e) What is the magnitude of the attractive force acting between the +1200 stat-coulombs charge and the −400 stat-coulombs charge?

6. Two small metal spheres A and B have radii of 5 cm and 10 cm respectively. Sphere A is charged with +100 stat-coulombs and sphere B is charged with +400 stat-coulombs. After charging, they are connected together by a long, thin metal wire. (a) Before they were connected, what was the absolute potential of sphere B? (b) After they were connected, what was the total capacitance of the combination? (c) After they were connected, what was the absolute potential of both spheres? (d) After they were connected, how much energy in ergs was stored in the combined spheres?

7. A simple condenser is made of 2 brass plates, each of 20 sq cm area placed parallel and .2 cm apart with air between them. What is the capacitance of this condenser?

8. Two condensers, each made of 2 metal plates 20 sq cm in area, .2 cm apart, are in parallel. One of them has air between its plates, and the other has glass whose dielectric constant is 5. (a) Deter-

mine the capacitance of each of the condensers. (b) What is the total capacitance of the combination? (c) If the potential difference between the plates of the air condenser is 2 stat-volts, what is the magnitude of its charge? (d) What is the total charge on the combination of condensers?

9. Two condensers, each having a capacitance of 4 stat-farads, are connected in series in a circuit to a battery which produces a potential difference at the terminals of the combination of 2 stat-volts. (a) What is the combined capacitance of the two condensers? (b) What is the magnitude of the charge on the condenser plates? (c) What is the potential difference between the plates of one of the condensers?

10. An isolated metal sphere is 4 cm in radius. It is charged until its absolute potential is 1.2 stat-volts. (a) What is the capacitance of this sphere? (b) What is the magnitude of the charge on the sphere? (c) How much energy was required to charge the sphere to a potential of 1.2 stat-volts? (d) If this charged sphere is connected by a thin wire to an uncharged sphere of 2 cm radius, what will be the total charge on both spheres? (e) After they are connected, what will be the difference in potential between the two spheres?

11. A sphere of 20 cm radius is charged to a potential of 2000 stat-volts. (a) What is the capacitance of the sphere? (b) How much charge does the sphere have? (c) How much work was done in charging the sphere? (d) What is the field intensity at a distance of 100 cm in air, due to the sphere?

12. What is the potential at the center of a cube 1 cm on a side if charges of $+100$ stat-coulombs are placed at each of the eight corners?

13. If, in the preceding problem, negative charges of 100 stat-coulombs are substituted for the positive charges on every other corner, what will be the potential at the center of the cube?

14. Two condensers, of 5 and 8 farads capacitance respectively, are connected in series. What is the capacitance of the combination?

15. A point charge has a magnitude of $+9 \times 10^{-9}$ coulombs. Two points A and B are 10^{-1} m and 20×10^{-1} m away respectively from the point charge. What is the difference in potential between the points A and B?

16. A capacitor has a potential difference between its plates of 200 volts. If its capacitance is .1 microfarad, what is the charge on each plate of the capacitor?

13

Electric Current

Certain substances, particularly metals, are better conductors of electricity than other substances. If a conductive path is provided between two points originally at different potentials, charge will move (flow). *Current* is defined as the flow of charge, and *intensity* of *current* (I) is defined as the time rate of flow of charge, i.e.:

$$I = \frac{Q}{t}$$

The direction of current is a matter of convention. Since the concept of positive charge is basic to electrical terminology, the conventional direction of current is from positive + to negative − in a simple conductor, i.e., one which does not include any source of potential rise, such as a battery. On the other hand, it is well known that current in metallic conductors is largely a matter of electron flow. Electrons are negatively charged particles, which therefore flow in the opposite direction from positive charges. Therefore, electron flow is from negative − to positive +. It is assumed that flow of positive charge in one direction is equivalent to the flow of negative charge (electrons) in the opposite direction.

The function of a battery is to develop and maintain the difference in potential necessary to move charges in a circuit consisting of the battery and an external conducting path which offers resistance to the flow. An analogous situation would be that of a hydraulic pump connected in a closed circuit of pipes through which a fluid flows as a result of the push of the pump. Electrically, the battery develops *electromotive force* (E), by chemical action or by some other method of converting energy to electrical energy from some other form, and the circuit displays *external resistance* (R_0) as well as *internal battery resistance* (b).

Ohm's Law states that in a series circuit such as shown in Fig. 72:

$$I = \frac{E}{R_0 + b}$$

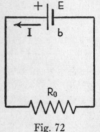

Fig. 72

The practical unit of current is the *ampere*, the practical unit of electromotive force is the *volt* (1 stat-volt = 300 volts), and the practical unit of resistance is the *ohm*. These are also the mks units of current, electromotive force, and resistance, respectively.

Comment on Units. Attention was called earlier to the competition between the metric cgs and mks units in electricity, and the increasing appeal of the latter to those persons dealing with the practical side of the subject. Actually the situation has been complicated by the development and acceptance over the years of *three* different sets of electrical units. One of these is the so-called *absolute cgs electrostatic system* based on the supposititious unit positive charge (stat-coulomb). The second is the so-called *absolute cgs electromagnetic system* based on the supposititious unit north pole of magnetism because of a magnetic effect dependent upon charge in motion (electric current). The third is the so-called *practical system* based on standardizable measurable quantities such as the ampere (defined empirically in terms of a certain chemical effect of current) and the volt (defined in terms of the properties of a certain chemical battery); or the ohm (defined in terms of the resistance of a certain filament of mercury). Interestingly, with the development of mks units in mechanics, it was discovered that by the use of an arbitrary constant ($k = 1/4\pi\varepsilon_0$) in Coulomb's Law and by redefining the ampere in terms of a different magnetic effect, all the units in the so-called practical system become the appropriate units in the mks system for electricity. Thus the mks system actually becomes a fourth system which is able to replace the other three and simplify the subject, at least as far as electric current phenomena are concerned.

Obviously the reason that the mks system of units has been so readily adopted in electricity is that it is essentially the same system that has always been used, with simply a change in name, in dealing with electric current phenomena. In electrostatics the absolute cgs electrostatic system still has some appeal.

Effects of Electric Current. Most of what is known about the electric current is the result of indirect evidence, since moving charge is an abstract concept not capable of direct sense perception. There are three major effects which have contributed most of the information accumulated concerning the electric current. These are (1) the heating effect, (2) the chemical effect, and (3) the magnetic effect. These will be considered in turn.

Heating Effect of Current. Joule discovered that the rate of heating varies directly as the second power of the current, i.e.

$$\frac{H}{t} \sim I^2 \quad \text{or} \quad \frac{H}{t} = RI^2$$

where R (resistance) is the proportionality factor. Recall that heat (H) is a form of energy, and that JH (J = the mechanical equivalent of heat or 4.18 joules per calorie) expresses heat in energy units (joules).

$$\therefore \frac{JH}{t} = \frac{W}{t} = RI^2 = P \ (\text{watts, or joules/sec})$$

where P represents the power dissipation. This is known as *Joule's Law*.

On the other hand, Ohm discovered (as noted already) that the intensity I of current in a simple series circuit consisting of a battery of electromotive force E and an external conductor which offers opposition to the flow (resistance R_0) is given by the expression

$$I = \frac{E}{R} = \frac{E}{R_0 + b}$$

where R is the total resistance of the circuit, including the external resistance R_0 and the internal battery resistance b.

Applied to a portion of a circuit which does not contain a seat of electromotive force (such as a battery provides) *Ohm's Law* states that the intensity of the current is given by

$$I = \frac{V}{R}$$

where V is the difference in potential (same as potential drop) across the resistor R.

Concept of Resistance. The concept of resistance is thus common to Joule's Law and Ohm's Law. In the former it suggests a characteristic of a given conductor by virtue of which heat is generated as current is maintained. In the latter it suggests that a given electrical conductor offers opposition to the flow of charge, i.e., current. In both cases, it suggests a physical characteristic of matter. It is also found that the resistance of a conductor depends upon the length L of the conductor, and inversely upon the area of cross section A of it. It also depends upon the material of which it is made. Thus

$$R = k \frac{L}{A}$$

where k is called the specific resistance or the resistivity of the material, i.e., the resistance of a given material having unit length and unit cross-sectional area.

Resistance is also found to depend upon temperature in the following manner:

$$R = R_0(1 + \alpha t) \quad \text{or} \quad \alpha = \frac{R - R_0}{R_0 t}$$

where R_0 is the resistance at $0°$ C and α is the temperature coefficient of resistance.

Resistances combined as in Fig. 73 are said to be connected in *series*. In this case (*Series*):

$$R = R_1 + R_2 + R_3$$

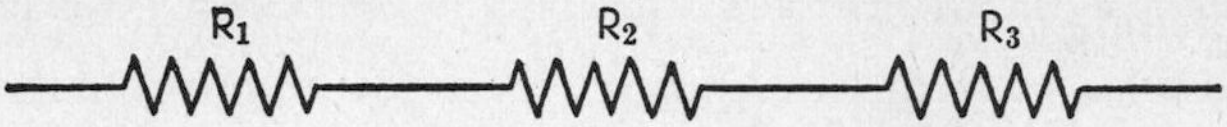

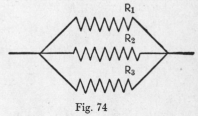

Fig. 73

Resistances combined as in Fig. 74 are said to be connected in *parallel*.

In this case (*Parallel*):

$$\frac{1}{R} = \frac{1}{R_1} + \frac{1}{R_2} + \frac{1}{R_3}$$

Fig. 74

The practical unit of resistance is the *ohm*. It is the resistance offered by a filament of liquid mercury 106.30 cm long, with constant cross section of one square millimeter, maintained at the temperature of melting ice ($0°$ C).

Chemical Effect of Current. Certain chemical solutions are capable of passing electric charges, i.e., of carrying current, when a battery is connected across two electrodes which dip into the solution. These solutions are called *electrolytes*. They contain positive and negative *ions,* which are the small particles that carry the charge. The electrode connected to the *positive* terminal of the battery is called the *anode,* and is the electrode at which the current is said to *enter* the electrolyte. The electrode connected to the *negative* terminal of the battery is called the *cathode,* and is the electrode at which the current is said to *leave* the electrolyte. The negative ion is called the *anion* because it is drawn to the anode; the positive ion is called the *cation* because it is drawn to the cathode.

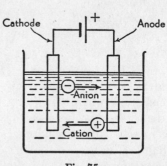

Fig. 75

Metallic substances in solution, and hydrogen, having positive valence, form positive ions; nonmetals and acid radicals, having negative valence, form negative ions. Thus metallic ions are attracted toward the cathode of an electrolyte cell, where their charges are neutralized and they collect. This is the basis of electroplating.

Faraday's Law of Electrolysis. Faraday discovered that the mass of material deposited at the cathode depends upon the intensity I of the current, the so-called electrochemical equivalent Z, and the time t. Thus

$$M = ZIt = ZQ$$

where $Q = It$.

Moreover $$Z = \frac{\text{atomic weight}}{\text{valence} \times 96{,}540}$$

In other words, 96,540 coulombs (approx) of electric charge will deposit one gram equivalent of a substance on the cathode, while causing an equal amount to go into solution at the anode.

$$96{,}540 \text{ coulombs} = 1 \text{ faraday}$$

The International Ampere. The chemical effect of current provides the legal unit of current. One *ampere* is that steady current which in one second will deposit .001118 g of silver from a silver solution. The ampere will be redefined later as an mks unit. Thus, legally, the coulomb is one ampere second.

From Ohm's Law, one legal *volt* is merely the product of one *ampere* times one *ohm*.

Problem Procedures

Problems in electric current phenomena are frequently directed toward analyzing electric circuits, to determine the intensity of current *in*, the potential differences *across*, and the amount of resistance *of* various portions of a circuit. Thus such problems involve applications of Ohm's Law, Joule's Law, and the concepts associated with resistance and electromotive force.

As in the case of problems in mechanics, there is a logical general approach to problems in electrical circuits. Many of the steps should look familiar to the student who has followed the presentation up to this point.

In the first place, *always* draw a diagram of the situation. This usually means a circuit diagram. Label every component of the circuit, including the unknown quantities as well as the known quantities, specifying just what it is that is to be determined. A typical problem might involve a circuit such as shown in Fig. 76. The problem might be to determine I, V_{AB}, I_3, and V_{CA}.

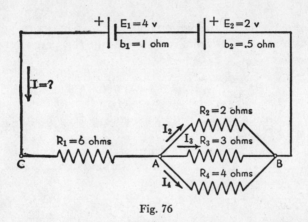

Fig. 76

The next step is to consider possible combinations of electromotive forces and possible combinations of resistances in the hope that the circuit may be simplified so that Ohm's Law can be utilized.

In the above example, it is clear that the two batteries are in series, whence a single electromotive force equal to the *algebraic* sum of the two can be substituted for both of them.

$$E = E_1 + E_2 = 4 - 2 = 2 \text{ volts}$$

(Note that E_2 is negative with respect to E_1, i.e., the batteries buck each other.) Moreover, this equivalent battery has a resistance:

$$b = b_1 + b_2 = 1 + .5 = 1.5 \text{ ohms}$$

It is also evident that the three resistances R_2, R_3, R_4 represent a parallel combination whose combined resistance is given by

$$\frac{1}{R_x} = \frac{1}{R_2} + \frac{1}{R_3} + \frac{1}{R_4}$$
$$= \tfrac{1}{2} + \tfrac{1}{3} + \tfrac{1}{4}$$
$$= .50 + .33 + .25 = 1.08$$
$$\therefore \quad R_x = \frac{1}{1.08} = .93 \text{ ohms}$$

Moreover, this equivalent resistance R_x is in series with R_1. Therefore all the external resistance in the circuit can be expressed as

$$R_0 = R_1 + R_x = 6 + .93 = 6.93 \text{ ohms}$$

Thus the entire circuit can be replaced by a simple equivalent circuit as in Fig. 77.

The third step is now to apply Ohm's Law to this simple circuit.

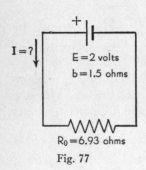

$I = ?$
$E = 2$ volts
$b = 1.5$ ohms
$R_0 = 6.93$ ohms

Fig. 77

$$I = \frac{E}{R_0 + b} = \frac{2}{6.93 + 1.50}$$
$$= \frac{2}{8.43} = .24 \text{ amperes} \quad Ans.$$

The fourth step is to consider by Ohm's Law again the potential drop across various parts of the circuit. Thus from C to A across R_1:

$$V_{CA} = IR_1 = .24(6) = 1.44 \text{ volts} \quad Ans.$$
and
$$V_{AB} = IR_x = .24(.93) = .22 \text{ volts} \quad Ans.$$

Again, by Ohm's Law, the current I_3 is given by

$$I = \frac{V}{R} \qquad \therefore \quad I_3 = \frac{V_{AB}}{R_3} = \frac{.22}{3} = .073 \text{ ampere} \quad Ans.$$

In addition to these answers, the potential drop across the

entire external circuit can be determined by Ohm's Law.

$$V_{CB} = IR_0 = .24(6.93) = 1.66 \text{ volts}$$

Now the potential drop V_{CB} is not only the potential drop across the external resistance R_0; it is also the potential drop across the battery. But by Ohm's Law

$$I = \frac{E}{R_0 + b} \qquad \therefore \quad IR_0 + Ib = E$$

or
$$V + Ib = E \qquad \therefore \quad V = E - Ib$$

By substituting values

$$V = 2 - (.24)(1.5)$$
$$= 2 - .36 = 1.64^+ \text{ volts}$$

which is, within the precision of the calculations, equal to the potential drop V_{CB} above.

Problem

Compare the cost of operating 3 lamps in series and in parallel on a 115 volt circuit, if each lamp has a resistance of 100 ohms. What is the power consumption in each case, and how many calories of heat are generated in each case during a one hour period?

Solution

Draw a diagram for each case (the symbol ω is often used to indicate "ohms").

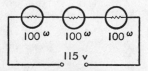 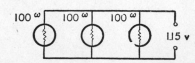

Fig. 78

By Joule's Law $\qquad P = \dfrac{JH}{t} = I^2R$

But $IR = V$.

$$\therefore \quad P = I \times IR = IV \quad \text{where} \quad I = \frac{V}{R} \text{ (Ohm's Law)}$$

In Series	*In Parallel*
$R = R_1 + R_2 + R_3 = 100 +$ $100 + 100 = 300$ ohms	$\frac{1}{R} = \frac{1}{R_1} + \frac{1}{R_2} + \frac{1}{R_3} = \frac{3}{100}$
$\therefore \quad I = \frac{115}{300} = .383$ ampere	$\therefore \quad R = 33.3$ ohms
whence $\quad P = IV = .383(115)$ $= 44.0$ watts $\quad$ *Ans.*	$I = \frac{115}{33.3} = 3.45$ amperes
	$P = 3.45(115) = 397.0$ watts $\quad$ *Ans.*

$$\therefore \quad \frac{\text{Power (Parallel)}}{\text{Power (Series)}} = \frac{397}{44.0} = 9 \text{ times} \quad Ans.$$

The heat generated is given by

$$JH = I^2Rt \quad \text{or} \quad H = \frac{I^2RT}{J}$$

In Series	*In Parallel*
$H = \frac{(.383)^2(300)(3600)}{4.18}$	$H = \frac{(3.45)^2(33.3)(3600)}{4.18}$
$= \frac{.147 \times 3 \times 3.6 \times 10^5}{4.18}$	$= \frac{11.9 \times 3.33 \times 3.6 \times 10^4}{4.18}$
$= .380 \times 10^5 = 38,000$ cal *Ans.*	$= 34.1 \times 10^4 = 341,000$ cal *Ans.*

Problem

How long will it take to silver plate 2.00 g of silver on a spoon, if the current in the plating bath circuit is maintained at 5 amperes?

Solution

This problem is obviously a direct application of Faraday's Law of electrolysis:

$$M = ZIt$$

where Z for silver is found from a handbook to be .001118 g/coulomb.

Thus $\qquad 2.00 = .001118(5)t$

$$t = \frac{2.00}{.005590} = \left. \begin{array}{l} 357.7 \text{ sec or} \\ 5.96 \text{ min} \end{array} \right\} \quad Ans.$$

Problem

The total resistance of a wire wound rheostat when "cold" is 300 ohms. In use it experiences a rise in temperature. How does the current which it draws on a 100 volt line after its temperature has risen 50° C compare with that drawn at the start when it is "cold"?

Solution

Recognize this problem as one involving a combination of Ohm's Law and the dependence of resistance upon temperature.

$$I = \frac{V}{R} \text{ (Ohm's Law)}$$

But $R = R_0(1 + \alpha t)$ where α for metals is .0038 °C (approx)
When "cold" $R_1 = 300$ ohms.

$$\therefore \quad I_1 = \frac{V}{R_1} = \frac{100}{300} = .333 \text{ ampere}$$

But when "hot" $R_2 = 300[1 + .0038(50)]$
or $R_2 - 300 = 300(.0038)(50)$
$$= 3 \times 3.8 \times 5 \times 10^0 = 57$$
$$\therefore \quad R_2 = 357$$

And $I_2 = \dfrac{V}{R_2} = \dfrac{100}{357} = .280$ ampere

Thus $I_2 = \dfrac{.280}{.333} I_1 = .84\, I_1$ (approx)   *Ans.*

The Wheatstone Bridge Circuit. One of the famous electric circuits in Physics is the so-called Wheatstone Bridge. It provides a method of measuring resistance, utilizing the concepts of potential drop and divided currents. If a circuit is arranged as in Fig. 79, it will be noted that at the branch point A, current I is divided into two branches I_1 and I_3. Similarly at D, the branch

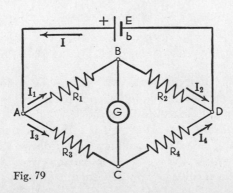

Fig. 79

currents I_2 and I_4 unite to give I. Moreover, if the point C is at the same potential as point B, there will be no flow registered by the galvanometer G, whereupon

$$I_1 = I_2 \quad \text{and} \quad I_3 = I_4$$

and also $\qquad V_{AB} = V_{AC} \quad \text{while} \quad V_{BD} = V_{CD}$

But $V_{AB} = I_1R_1$; $V_{AC} = I_3R_3$; $V_{BD} = I_2R_2$; and $V_{CD} = I_4R_4$.

$$\therefore \quad I_1R_1 = I_3R_3$$

and $\qquad\qquad\qquad I_2R_2 = I_4R_4$

whence $R_1 = (R_3/R_4)R_2$ if the galvanometer registers zero.

Since only the ratio R_3/R_4 rather than the actual values of R_3 and R_4 are involved in this expression, the resistances R_3 and R_4 can be replaced in the circuit by a single wire as in Fig. 80.

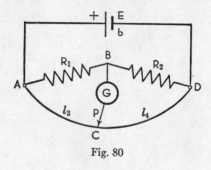

Fig. 80

To use the Wheatstone Bridge, the pointer P is slid along the wire to locate the point C at which the galvanometer G reads zero, whence

$$R_1 = \frac{l_3}{l_4} R_2$$

Thus the resistance of an "unknown" resistor R_1 is determined in terms of two measurable lengths l_3 and l_4, and a laboratory standard resistor whose resistance is R_2.

Problem Possibilities. Although the Wheatstone Bridge is a laboratory instrument which enables one to determine the value of an "unknown" resistance by comparison with a laboratory standard resistance, it can be used to illustrate applications of

Ohm's Law in classroom problems. Using the first of the two diagrams, students can be expected to determine I_2 if I_1 is given for a "balanced" bridge, or to determine the value of the drop in potential V_{AB} from point A to B if the current I_1 is known, or the drop in potential from point B to C if the bridge is "balanced," etc.

Problems: Electric Currents

1. The potential difference between points A and B is 2 volts. The current from point A to B is .4 ampere. What is the magnitude of the resistance?

2. The potential difference between the ends of a 20 ohm resistance is 5 volts. What is the magnitude of the current in the resistance?

3. Two resistances of 5 ohms and 10 ohms are in series, and the current in the 5 ohm resistance is $\frac{1}{2}$ ampere. What is the potential difference between the ends of the 10 ohm resistance?

4. A steady electric current of .25 ampere is established in a conductor of 120 ohms. What is the potential difference between the ends of this conductor?

5. A dry cell whose emf is 1.5 volts and whose internal resistance is .2 ohm is in series with a resistance of 4.8 ohms. What is the magnitude of the current in the circuit?

6. The current in a wire remains at 5 amperes for 50 sec. What is the total charge which flows during this interval?

7. In a certain circuit the current in a 5 ohm resistor is 15 amperes. What is the potential difference across the terminals of the resistor?

8. A 5 ohm resistor carries a current of 10 amperes for a period of 6 sec. At what rate, in watts, is heat being generated in the resistor?

9. The resistivity of a certain metal is 2.5×10^{-6} ohm cm. How long must a piece of wire .01 cm² in cross section be for its resistance to be 5 ohms?

10. The electrochemical equivalent of silver is .001118 g/coulomb. How many grams will be deposited if a current of 2 amperes passes through a silver nitrate solution for 120 sec?

11. A 10 ohm resistor carries a current of 5 amperes. How much energy, in joules, will be dissipated in the form of heat during a period of 6 sec?

12. The current in a lamp of 300 ohms resistance is .4 ampere. What is the power rating of the lamp in watts?

13. A battery having an emf of 4.8 volts and an internal resistance of .3 ohm supplies 2.4 amperes to an external circuit. What is the resistance of the external circuit?

14. The emf of a battery is 7.5 volts on open circuit. The voltage across the terminals drops to 7.2 volts when delivering 3 amperes to the external resistance. (a) What is the internal resistance of the battery? (b) What is the external resistance of the circuit? (c) What is the power loss in the battery? (d) What is the power expended in the external circuit?

15. Four identical resistances are in series. The potential difference between opposite ends of the combination is 10 volts, and the current in the leads is .5 ampere. (a) What is the current in each resistance? (b) What is the potential drop across each resistance? (c) What is the value of each individual resistance? (d) What is the resistance of the whole combination?

16. A storage battery delivers 2 amperes for a period of 50 hours to a resistor. What is the total energy supplied during this time to the resistor if the potential difference across the resistor is 6.0 volts?

17. A dry cell, emf = 1.5 volts, internal resistance = .1 ohm, is connected to an external resistance $R = .4$ ohm. (a) What is the current in R? (b) What is the current in the cell? (c) What is the potential difference across R? (d) What is the potential difference across the cell terminals? (e) How much work is done by the cell in sending 1 coulomb once around the entire circuit? (f) At what rate, in watts, is energy being dissipated as heat by the resistor? (g) At what rate, in watts, is energy being dissipated as heat within the cell?

18. Two resistances, of 10 ohms each, dissipate 600 watts each when connected in series to a certain generator. What will be the power dissipated by each resistor when they are connected in parallel to the generator? (Assume the potential difference across the generator terminals to be the same in each case.)

19. A storage battery, emf = 6.0 volts, internal resistance = .01 ohm, is connected to an external resistance $R = .01$ ohm. (a) What is the current in the battery? (b) What is the potential difference across R? (c) What is the potential difference across the battery terminals? (d) How much work is done by the battery in sending 10 coulombs once around the entire circuit? (e) At what rate, in watts, is heat dissipated by the resistor? (f) If the resistor is disconnected, what is the potential difference across the battery terminals? (g) What is the current in the battery with the resistor disconnected?

20. How much does it cost to operate a 1000 watt light bulb for 10 hours if electrical energy costs 5 cents per kilowatt hour?

21. A circuit consists of a battery, having electromotive force $E = 24$ volts and internal resistance $b = 2$ ohms, connected in series

with a parallel resistance combination consisting of three 9 ohm resistors. What is the current in each of the 9 ohm resistors?

22. A certain set of colored Christmas tree lights consists of 7 lamps connected in parallel and is designed for use on a 115 volt line. If each lamp is rated at 15 watts, (a) how much current does each draw? (b) how much current does the combination draw? (c) what is the resistance of each lamp?

23. In a certain cottage one of the electrical circuits carries in parallel four 60 watt lamps, two 100 watt lamps, and an outlet into which is connected an appliance which draws 5 amperes, all on the 115 volt line. If electrical energy costs eight cents per kilowatt hour, how much does it cost to operate these lights and this appliance for 12 hours?

24. Referring to the circuit of the preceding problem, (a) what is the potential drop across the appliance? (b) what is the total current drawn? (c) what is the current in each 100 watt light? (d) what is the hot resistance of each 60 watt light?

25. A cell has an emf of 1.6 volts, and an internal resistance of .2 ohm. If it is connected across a 15.8 ohm resistor, what is the potential drop across the battery itself?

26. A combination of resistors consists of 3 ohms, 4 ohms, and 6 ohms all in parallel. This combination is connected in series with a .67 ohm resistor and a battery having an emf of 6 volts and an internal resistance of 1 ohm. (a) What is the potential drop across the parallel resistance combination? (b) What is the current in the 4 ohm coil?

27. How long will it take to deposit .5 g of copper on the cathode of an electrochemical cell at a current intensity of 5 amperes? The electrochemical equivalent of copper may be taken as .00033.

28. An electromotive force of .01 volt due to thermoelectricity is impressed across a bus bar of copper 10 cm long and 1 sq cm in cross section. How much heat is developed in 10 min? Take the resistivity of copper to be 1.7×10^{-6} ohm cm.

29. An electric immersion heater which draws 5 amperes operates for 20 min in 2000 g of water contained in a thermally insulated jar originally at 19° C. The resistance of the heater is 4 ohms. What rise in temperature will take place?

30. Two batteries whose emf are 12 volts and 6 volts and whose internal resistances are 1 and .5 ohms respectively, are connected so as to oppose one another in series with a divided circuit of two branches a and b. The current in the batteries is 3 amperes, while the current in a is 2 amperes. What is the drop in potential across b, and what are the resistances of a and b?

14

Magnetism and Magnetic Effects of Currents

Before studying the magnetic effect of current it is well to consider the concepts of magnetism. A bar magnet displays the property of attracting to itself particles of iron, such as iron filings, and the property of orienting itself in a north-south direction when suspended by a string attached to its middle as in Fig. 81. These so-called magnetic effects historically have been attributed to the existence of so-called magnetic poles located near the ends of the magnet, the one being called north, and the other being called south. These poles obey Coulomb's Law of magnetism, which is identical in form with Coulomb's Law of electric charge. Like the unit positive charge concept in electricity, the supposititious unit north-pole concept in magnetism has provided the basis for the development of magnetic terminology and a quantitative study of the subject. For this reason magnetic poles will be discussed briefly in this text as background material for the more sophisticated modern treatment of magnetism which will follow.

Coulomb's Law of Magnetism. (a) Like poles repel and unlike poles attract. (b) The force of attraction or repulsion between magnetic poles depends directly upon the product of the pole strengths and inversely upon the second power of the separation. It also depends upon the medium between the poles.

$$F = \frac{mm'}{\mu r^2}$$

where m and m' represent the magnitudes of the pole strengths, r is the separation, and μ is the magnetic permeability of the medium.

As in the case of electricity, where the supposititious unit of charge is defined in terms of Coulomb's Law of electricity, the *unit north pole* of magnetism is defined in terms of Coulomb's Law of magnetism. It is defined as that unit of pole strength such that two identical poles 1 cm apart in vacuum will repel each other with a force of 1 dyne.

Magnetic Field Intensity. The force per unit north pole at a point is referred to as the *magnetic field intensity H* at that point. It is a vector quantity just as is its counterpart in static electricity, the electric field intensity E.

$$H = \frac{m}{\mu r^2}$$

at a distance r from a pole of strength m in a medium whose permeability is μ. It is expressed in dynes per unit pole or in *oersteds*.

Magnetic Moment. The *magnetic moment M* of a bar magnet whose pole strength is m and whose length is L is given by

$$M = mL$$

Magnetic Lines of Force. As in electrostatics, the line of force concept is used to represent graphically the field intensity distribution in the vicinity of a magnet, or the intensity distribution of any magnetic field. By convention, lines are drawn pointing in the direction that a supposititious unit north pole would move if free to do so at each point in question, and in such numbers per square centimeter as the number of dynes per unit pole, or oersteds of field intensity, indicate at each place. For a bar magnet they appear as in Fig. 82. At a point such as P the number of lines crossing perpendicularly an area of one square centimeter designates the field intensity at that point in oersteds.

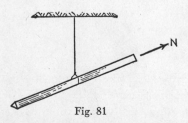

Fig. 81

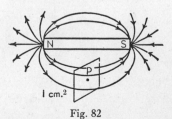

Fig. 82

Magnetic Flux Density. If instead of drawing lines according to values of H as above, we draw them according to values of μH, we note that such lines are continuous. They not only seem to emanate from the north pole and extend in air around to the south pole, but they also can be thought of as passing through the magnet from the south pole to the north pole. Such lines are called flux lines, and the number per square centimeter at a given place is called the flux density B. The ratio of B to H is the permeability μ.

$$\mu = \frac{B}{H}$$

B is expressed in lines per cm², or in gauss.

A substance like iron, when placed in a magnetic field H, becomes magnetized in such a way that it exhibits more flux lines per unit area than existed in the absence of the iron. Magnetism is said to be induced under these circumstances. A substance is said to be *paramagnetic* when it thus concentrates lines and *diagmagnetic* when it disperses them. Iron, cobalt, and nickel are exceedingly paramagnetic and are said to be *ferromagnetic*.

Terrestrial Magnetism. The earth is assumed to be a large spherical magnet displaying two poles, one near Hudson Bay in Canada and the other one in Antarctica. The fact that these poles do not coincide with the geographic poles leads to the concept of the *angle of declination*, which is the angle between the geographic meridian and the magnetic meridian at a given location.

The fact that the earth is not flat gives rise to the concept of the *angle of dip*. At all places except along the magnetic equator of the earth the compass needle dips if allowed to do so, making an angle with the horizontal called the angle of dip. Ordinarily a compass needle is weighted to offset the tendency to dip so that it remains horizontal. This means that the earth's magnetic field has a vertical as well as a horizontal component such that

$$\tan \theta = \frac{H_v}{H_h}$$

where θ is the angle of dip, H_v is the vertical component and H_h is the horizontal component of the field intensity.

Comment. The student is again emphatically reminded that

the magnetic pole is a supposititious unit like the unit charge in electricity. Whereas a naturally occurring unit of electricity, the electron, has been discovered, no real magnetic pole has been, or probably ever will be, discovered because the electron in motion, i.e., the electric current, displays its own magnetic effect. The real value of the pole concept is the basis which it has provided in the development of magnetic terminology. Although today it is generally regarded as an obsolete concept, its appreciation by the beginning student is felt to help him understand what it is that the more elegant modern treatment attempts to explain.

Problem Procedures

The real purpose of problems in magnetism is to clarify magnetic concepts. With this in mind, it should be clear to the student that the best approach to these problems, as indeed, by now it should be clear, to all problems in Physics, is to understand thoroughly the definitions of all the concepts involved. Always start with the defining equation and see where it leads. The basic law here is the law of Coulomb:

$$F = \frac{mm'}{\mu r^2}$$

By virtue of the definition of the unit north pole, cgs metric units are invariably used in magnetostatics when the pole concept is used.

Problem

A magnetic south pole of 15 units strength, when placed 10 cm away from another pole in air, experiences a repulsion of 300 dynes. What is the nature and strength of the second pole?

Solution

Obviously, since the force is one of repulsion, the second pole is a like pole, i.e., a south pole. *Ans.*

Obviously, also, the second answer is found from Coulomb's Law $F = mm'/\mu r^2$ where $F = 300$ dynes, m is 15 poles, $\mu = 1$ for air, and $r = 10$ cm.

$$\therefore \quad m' = \frac{\mu r^2 F}{m} = \frac{1(10)^2(300)}{15} = 2000 \text{ poles} \quad Ans.$$

Problem

What is the magnetic field intensity at a point in air 30 cm away from and on the extension of the line of a bar magnet 10 cm long whose magnetic moment is 400 poles cm, if the point in question is nearer the north pole of the magnet?

Solution

Draw a diagram.

Fig. 83

Recall that field intensity H is a vector quantity, which with respect to a pole strength m is given by the expression

$$H = \frac{m}{\mu r^2}$$ which follows directly from Coulomb's Law $F = \frac{mm'}{\mu r^2}$

where
$$H = \frac{F}{m'} = \frac{m}{\mu r^2}$$

In this problem, the pole strength can be evaluated because the magnetic moment is given by the defining equation $M = mL$, where M is 400 poles cm, and L is 10 cm.

$$\therefore \quad m = \frac{M}{L} = \frac{400}{10} = 40 \text{ poles}$$

It follows that due to the north pole of the magnet, the field intensity H_1 at P is

$$H_1 = \frac{m}{\mu r^2} = \frac{40}{1(30)^2} = \frac{40}{900} = .044 \text{ oersteds to the right}$$

Also, the field intensity H_2 due to the south pole is

$$H_2 = \frac{40}{1(40)^2} = \frac{1}{40} = .025 \text{ oersteds to the left}$$

$$\therefore \quad H \text{ is the vector sum of } H_1 \text{ and } H_2$$

or $\qquad H = \overrightarrow{.044} - \overleftarrow{.025} = .019 \text{ oersteds to the right} \quad Ans.$

Problems: Magnetism

1. A magnetic pole placed in a field of 10 oersted intensity experiences a force of 15 dynes. Find the pole strength.
2. A bar magnet has a magnetic moment of 120 poles cm and a length of 12 cm. Find the pole strength.
3. Two identical isolated magnetic poles are placed 50 cm apart in a vacuum and the force between them is found to be 4 dynes. What is the strength of these poles?
4. A bar of iron with a permeability of 500 develops a flux density of 1000 gauss when in a magnetic field of intensity H. Calculate H.
5. The true value of the earth's magnetic field intensity is .6 oersted and the angle of dip is 75°. Find the horizontal component of the earth's field (sin 75° = .97, cos 75° = .26, tan 75° = 3.73).
6. Two long narrow similar bar magnets are held with the two north poles 5 cm apart in air. If they repel with a force of 800 dynes, what is the strength of their poles, assuming them to be identical?
7. What is the intensity of the magnetic field at a point 10 cm from the center of a bar magnet 20 cm long, on a line perpendicular to the axis of the magnet if the pole strength is 400 poles? What is the moment of this magnet?
8. A bar magnet 20 cm long is made of iron of permeability 500. (a) If this magnet has a pole strength of 400 poles, what is the field intensity H (magnitude and direction) at the midpoint of the magnet due to the poles? (b) What is the flux density B at this point?
9. What is the angle of dip at a place where the magnitude of the vertical component of the earth's field is 3 times as great as that of the horizontal component?
10. What is the magnitude and direction of the earth's magnetic field at a place where the horizontal component is .22 oersted and the angle of dip is 70°?

The Magnetic Effect of Current. The Danish physicist Oersted discovered in 1819 that a compass needle was disturbed if it was placed near a current-carrying wire. Further experimentation shows that a conductor carrying an electric current always displays a magnetic field encircling it. Lines of force drawn to indicate the distribution of magnetic field intensity near a current-carrying wire are concentric circles. If such a wire is *imagined* to be grasped by the right hand with the thumb pointing in the direction of current flow (i.e., the direction of flow of positive charge), the fingers wrap around the wire in the direction of the field. See Fig. 84.

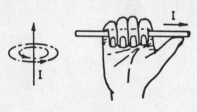

Fig. 84

This discovery was of far-reaching importance. It raised questions concerning the explanation of magnetic effects in terms of unit poles because it indicated that a magnetic field can be, and always is, produced by an electric current which is merely a stream of electrical charges. Indeed it suggested that one or the other of the two concepts, the unit positive charge of electricity or the unit north pole of magnetism, was unnecessary.

By now the existence of the electron as an electrical particle has been so completely established that the pole concept has become obsolete, and the explanation of magnetism has had to be updated to relate it to the electric current. The modern treatment will follow directly in this text, but to many it seems so sophisticated that it practically requires an appreciation of the earlier and perhaps more naive views in order to understand what it is that the newer treatment is attempting to explain. Part of the sophistication arises from the incorporation of the increasingly popular mks units, which require the introduction of certain arbitrary constants.

Modern Treatment of Magnetism. Up to this point magnetism has been considered from the historical viewpoint based on phenomena associated with the bar magnet and the compass needle. Consistent with the fundamental objectives of the science of Physics, however, to explain the most phenomena with the fewest possible assumptions, regardless of the mathematical complications which might be encountered, a somewhat different approach is required. Following the discoveries of Oersted and others, it is now believed that *all* of the so-called magnetic effects are due to moving electrical charges, including the magnetic properties of permanent (bar) magnets. The latter are explained by the circulation of electrons within the magnets.

The Magnetic Field. Electrostatic forces governed by Coulomb's Law exist between electric charges, but over and above such electrostatic forces, moving charges exert additional so-called *magnetic forces* on one another. These forces can be made evident by the deflection of a stream of electrons in a cathode ray tube, a device which is not as familiar as the bar magnet but is nevertheless fairly common in the present electronic age (television tube, for example). By definition, any region where such magnetic forces are experienced is called a *magnetic field.* It is characterized at each point by a vector quantity called *magnetic induction,* usually represented by the symbol B. (Had not the term "magnetic field intensity" been preempted for another pole-related concept, force per unit pole H, the latter would probably have been used instead of "magnetic induction" to characterize the magnetic field.) Magnetic induction turns out to have the same quantitative value at a given point in a field as the *magnetic flux density* concept encountered in the earlier pole discussion, so the symbol B is appropriately used to designate it.

Magnetic Induction B. The vector quantity B which characterizes a magnetic field is specified in the following manner. *First,* it is found by experience that when a charge q moves with velocity v in a magnetic field, a magnetic force F generally acts on it depending on the charge, the velocity, and the direction of the motion.

1. *Direction of B.* The above force F varies with the direction of v with respect to a given magnetic field. Consider a stream of electrons in a small cathode ray tube placed in a magnetic field. This stream will be deflected by the field in such a way as to make the force apparent. The amount of the deflection varies with the direction of the moving electrons (the orientation of the tube), there being a direction for which it is a maximum, and at right angles to which it is zero. This unique direction for which the deflection is zero in a given magnetic field is *defined* as the *direction of the magnetic field* and is therefore the direction of B.

2. *Magnitude of B.* The magnitude of the force is found to be

proportional to the charge q and the velocity component perpendicular to the direction of the field (i.e., $v \sin \phi$ where ϕ is the angle between the direction of v and the direction of B). B is the proportionality factor in this proportionality.

Thus $\qquad F = Bqv \sin \phi \qquad \therefore B = \dfrac{F}{qv \sin \phi}$

$$\left(\text{Expressed in } \frac{\text{newton}}{\text{coulomb meter/second}} \right)$$

Note! Mks units are usually used in the modern treatment of magnetism.

3. *Sense of B.* The sense of B as one of the three mutually perpendicular vectors F, B, and $v \sin \phi$ is established by convention, noting that the conventional direction of electric current is the direction taken by a moving *positive* charge rather than that of the electron (negative charge). Consider Fig. 85 which

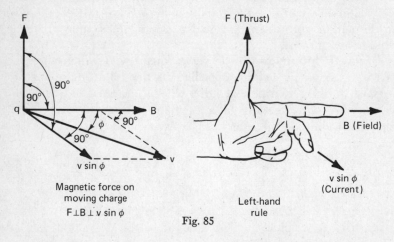

Magnetic force on
moving charge
$F \perp B \perp v \sin \phi$

Left-hand
rule

Fig. 85

suggests a so-called *left-hand rule*. The left-hand rule states that if the thumb and first two fingers of the left hand are held as in Fig. 85, the forefinger points in the direction of the field (magnetic induction B), the second finger points in the direction of the velocity component of positive charge perpendicular to B, and the thumb points in the direction of the magnetic force (side thrust).

Lines of Magnetic Flux. It is often convenient to represent magnetic fields graphically rather than by a tabulation of B values. See Fig. 82. Lines of magnetic flux are suggested which by their directions indicate the direction of the field at each point, and by their concentrations (the number of them drawn perpendicular to a unit of area) indicate the magnitude of the field at each point. Thus the *magnetic induction* becomes the same as the *flux density*. If flux is represented by lines (the mks unit is the weber) crossing an area A, the flux density

$$B = \frac{\Phi}{A} \text{ weber/meter}^2$$

In cgs units (when encountered) Φ is expressed in maxwells, and B in maxwells per sq cm (gauss).

Therefore, since

$$B = \frac{\Phi}{A} = \frac{F}{qv \sin \phi} \quad \therefore \quad \frac{1 \text{ weber}}{\text{meter}^2} = \frac{1 \text{ newton}}{\text{coulomb meter/sec}}$$
$$= \frac{1 \text{ newton}}{\text{ampere meter}}$$

Also $\qquad \dfrac{1 \text{ maxwell}}{\text{cm}^2} = 1 \text{ gauss} = 10^{-4} \dfrac{\text{weber}}{\text{meter}^2}$

where maxwell is the name given to one line/cm² and weber is the name given to 1 line/meter².

Magnetic Properties of Matter. Magnetic fields are set up by moving charges not only in air (or vacuum) but also in solid matter, where it is assumed that circulating electrons within matter are responsible for the magnetic field. Iron, cobalt, and nickel are the most conspicuous naturally occurring magnetic substances, but magnetic alloys can be manufactured. In such "permanent" magnets the extent to which the substance is magnetic compared to vacuum is designated by a characteristic called *permeability* (μ) which is somewhat analogous to the electrical dielectric constant concept (K), and which is expressed in weber/ampere meter. The ratio B/μ becomes a meaningful concept when associated with a magnetic substance in that this ratio characterizes both that part of a field due to the internal circulation of electrons and that part of it which may be due to external electric currents. This is the concept that was earlier referred to (pole theory) as the magnetic field intensity (H) near the poles (or ends) of a bar magnet; i.e., $\mu = B/H$ as before.

Force on a Conductor which Carries Charge across a Magnetic Field. The side thrust forces on individual electrons in a current-carrying conductor are transmitted to the conductor as a whole. For the case in which B, v, and F are mutually perpendicular (see diagram of left hand, Fig. 85), and recalling that the direction of current is the opposite of the direction of electron flow, the side thrust force is $F = Bqv = BLI$ for a conductor of length L, crossing a field B, and carrying a current I.

In *mks units:* F in newtons, B in weber/meter^2, L in meters, and I in amperes.

In *cgs units:* F in dynes, B in maxwell/cm^2, L in cm, but I must be expressed in a unit which is ten times as large as the ampere to balance the units. Such a unit of current is sometimes called the *ab-ampere*. 1 ab-ampere = 10 amperes.

Note! Before the advent of mks units, and on the pole theory of magnetism, the ab-ampere was the basis of a whole system of units which included the ab-volt, the ab-ohm, etc., and which competed with the stat-system (stat-coulomb, stat-volt, etc.), in addition to the so-called practical system (ampere, volt, ohm, etc.). The appeal of the mks units is largely that this system eliminates the others along with the attendant confusion.

Applications of Side Thrust. It is evident that if the current-carrying wire which is stretched across a fixed magnetic field as shown in cross section in Fig. 86a is movable, the side thrust acting on it will actually move it, as shown, in the direction of

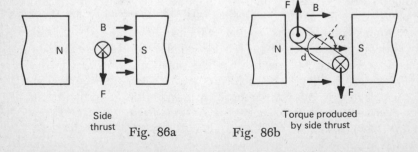

Side
thrust   Fig. 86a

Fig. 86b   Torque produced
by side thrust

the force, as determined by the left-hand rule (Fig. 85). The cross $\otimes$ is intended to indicate the tail or feathered end of an arrow pointing into the paper, i.e., the current points into the paper.

If a rectangular loop of wire, shown in cross section in Fig. 86b with length L (into the paper) and width d, is stretched as shown across a field B, the direction of the current in one leg of the loop will be the opposite of that in the other leg. Consequently the side thrust forces will develop a torque whose moment will tend to rotate the coil into a position across (perpendicular to) the field. It can be shown that whatever the shape of the loop with N turns, the torque τ can be expressed

$$\tau = NIAB \sin \alpha$$

if the current is I and the area of the loop is A. Such applications of side thrust are to be found in electric meters (ammeters, voltmeters, wattmeters, etc.) and electric motors, illustrating the practical importance of the magnetic effect of the electric current.

Evaluations of Magnetic Induction Due to Electric Currents. The preceding has dealt with the detection of magnetic fields resulting from moving charges and the quantitative specification of the forces which they exert on moving charges (currents). The Oersted experiment indicates that a magnetic field is generated by an electric current, but as yet no mention has been made of the *magnitude* of the magnetic induction as related to the intensity of the current producing it, although the *direction* of B is found always to be such as to *encircle the conductor* as in Fig. 84. Three different current-carrying conductor configurations will be considered.

1. *Field at Distance* d *from an Infinitely Long Straight Wire.* Fig. 87b₁.

$$B = k' \frac{2I}{d}$$

In cgs units $k' = $ unity, but I must be expressed in ab-amperes. In mks units $k' = 10^{-7}$ weber/ampere meter, but for convenience a new quantity μ_0 is defined such that $\mu_0 = 4\pi k' = 12.5 \times 10^{-7}$ weber/ampere meter.

$$k' = \frac{\mu_0}{4\pi} \quad \text{and} \quad B = \frac{\mu_0}{4\pi} \frac{2I}{d} \quad (I \text{ in amperes})$$

Biot and Savart discovered this relation experimentally.

2. *Field at Center of a Circular Turn of Radius R.* Fig. 87b$_2$.

$$B = \frac{\mu_0}{4\pi} \frac{2\pi I}{R}$$

3. *Field along the Axis of a Very Long Solenoid.* Fig. 87b$_3$.

$$B = \frac{\mu_0}{4\pi} \frac{4\pi NI}{L}$$

where N is the number of turns and L is the length of the solenoid.

The above are three special cases of a general formulation of magnetic induction B which is expressible only in terms of a differential equation called the Ampere-Laplace Rule. This treats the magnetic induction as made up of an infinite number of contributions, each contributed by an increment of length ds and indicated as follows.

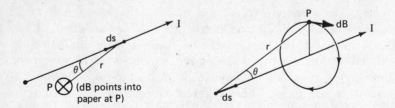

Two representations of Ampere LaPlace rule

Fig. 87a

With reference to Fig. 87a

$$dB = k' \frac{I\, ds}{r^2} \sin \theta = \frac{\mu_0}{4\pi} \frac{I ds}{r^2} \sin \theta$$

whereupon $\quad B = \frac{\mu_0}{4\pi} \int \frac{I \sin \theta}{r^2}\, ds \quad$ (a vector sum)

Special Cases

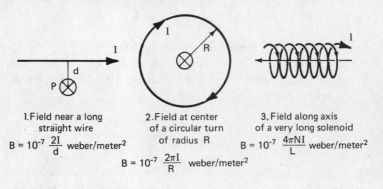

1. Field near a long straight wire

   $B = 10^{-7}\ \dfrac{2I}{d}$ weber/meter^2

2. Field at center of a circular turn of radius R

   $B = 10^{-7}\ \dfrac{2\pi I}{R}$ weber/meter^2

3. Field along axis of a very long solenoid

   $B = 10^{-7}\ \dfrac{4\pi NI}{L}$ weber/meter^2

Fig. 87b

Force between Parallel Conductors. A long current-carrying conductor produces a certain magnetic field at a distance d from itself. If a second current-carrying conductor lies parallel to the first one and is separated from it by a distance d, it produces a magnetic field where the first one is located. Each wire will experience a side thrust due to the current in the other one. If the two currents point in the *same direction* it can be shown that the mutual force is one of *attraction*, and vice versa. The magnitude of this force, per unit length of wire, can be shown to be

$$\frac{F}{L} = \frac{\mu_0}{4\pi} \frac{2II'}{d}$$

where I is the current in the one wire and I' is the current in the other one.

Definition of the Ampere. At long last the *ampere*, which is not only the unit of current in the mks system but is the basic electrical unit in that system, from which all the others are derived, is *defined* as that unvarying current which, if present in each of two parallel wires of infinite length and one meter apart in empty space, causes each conductor to experience an attractive force of exactly 2×10^{-7} newtons per meter of length, if the currents are directed the same way in each conductor. Thus 1 coulomb = 1 ampere second. This means that the amount of charge in the mks system of units is defined in terms of a force between *moving* charges rather than the electrostatic force between charges as in the cgs system of units.

Problem Procedures

Problems dealing with the Oersted discovery can be little more than numerical applications of the side-thrust phenomenon and of the Ampere-Laplace relationship:

$$\Delta B = 10^{-7} \frac{I \sin \theta}{r^2} \Delta s \ (\text{weber/meter}^2)$$

which can be handled only for those special cases for which the differential equation can be solved. Three special cases have been considered in this text.

With respect to units, the above formulation with the 10^{-7} factor implies mks units throughout, i.e., I in amperes, r and s in meters. Of course cgs units may be used, but if linear quantities are expressed in centimeters, I must be expressed in ab-amperes (1 ab-amp = 10 amp), B in gauss and the 10^{-7} factor becomes unity. Since the so-called ab-units are still in use, although the mks units seem to be rapidly replacing them, problems in this text will illustrate both systems.

Problem

How far from an infinitely long straight wire carrying 10 amperes must a point be located for the magnetic induction at that point due to the current to be 5×10^{-5} weber/meter^2?

Solution

This problem is a direct application of the Biot and Savart Law:

$$B = 10^{-7} \frac{2I}{d}$$

$$d = \frac{10^{-7}2I}{B} = \frac{10^{-7}20}{5 \times 10^{-5}} = 4 \times 10^{-2} \text{ meters} = 4 \text{ cm} \quad Ans.$$

Problem

What is the magnetic induction B at the center of a circular cable consisting of 100 turns of wire having a common radius of 20 cm carrying 15 amperes?

Solution

This problem obviously involves the relation $B = 2\pi NI/10r$, which is simply N times the value of the induction at the center of a single circular turn, where N is the number of turns.

$$\therefore \quad B = \frac{2\pi(100)15}{10(20)} = 15\pi \text{ gauss} \quad Ans.$$

Problem

Calculate the magnetic induction B along the axis of a very long 2.5×10^{-1} meters solenoid of 1000 turns of wire if the radius of the coil is 2×10^{-2} meters and the current in the wire is 15 amperes.

Solution

As shown above, for a long solenoid $B = 10^{-7} \dfrac{4\pi NI}{L}$ where $N = 1000$ turns, $I = 15$ amperes, and $L = 2.5 \times 10^{-1}$ meter.

$$B = \frac{10^{-7} \, 4\pi(1000)(15)}{2.5 \times 10^{-1}} = 7.54 \times 10^{-2}$$
$$= .0754 \text{ weber/meter}^2 \quad Ans.$$

Problem

If a wire carrying 2 amperes lies perpendicularly across a uniform magnetic field of flux density 5×10^{-2} weber/meter^2 in such a manner that 15 cm of the wire are subjected to the field (i.e., if the pole pieces of the magnet are 15 cm across), how much side thrust is experienced by the wire, and which way does it act?

Solution

By the left-hand rule, it is seen that the thrust is downward. *Ans.* The magnitude of the thrust is given by the side-thrust relation.

$$F = BLI$$
$$\therefore \quad F = 5 \times 10^{-2} \times 15 \times 10^{-2} \times 2 = 15 \times 10^{-3} = .015 \text{ newton}$$
$$Ans.$$

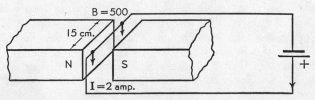

Fig. 88

Problems: Magnetic Effect of Current

1. A straight wire 100 cm long and carrying a current of 50 amperes is placed in a region where the magnetic flux density is uniform and equal to 25 gauss. Calculate the maximum force that could be exerted on the wire due to the flux.

2. A straight wire 10 cm long carrying a current of 50 amperes is placed in a region of uniform magnetic flux density whose direction is perpendicular to that of the wire and whose magnitude is 2.5×10^{-3} weber/meter^2. What is the magnitude of the force experienced by the wire because of the flux?

3. What is the magnitude of the magnetic induction at the center of a circular plane coil of wire with five turns and a radius of .2 m if the current in the wire is 2 amperes?

4. A long solenoid of 1000 turns is 50 cm long. A current of 20 amperes is established in the wire. (a) What is the magnitude of the flux density at the exact center of the solenoid? (b) If a core of iron whose permeability is 1200 is pushed into the solenoid, what does the flux density become at the center? (c) If the iron core has a cross-section area of 2 sq cm, what is the total flux emerging from the end of the core?

5. The field strength outside of a long straight wire is given by $H = 2I/d$. What is the magnitude of the force that will be experienced by a magnetic pole of strength 4 poles placed a distance of 10 cm from the wire if the current in the wire is 3 ab-amperes?

6. A tangent galvanometer, an early type of current-measuring instrument, is a device consisting of a vertical coil placed such that its plane is parallel to the north-south magnetic direction. At its center is placed a horizontal compass needle which, in the absence of a current in the coil, points along the plane of the coil. The earth's field has an induction value of 3×10^{-5} weber/meter^2, and the flux density B at the compass due to the coil current is 6×10^{-5} weber/meter^2. At what angle, with respect to the north, will the compass needle be deflected?

7. How much current is required in a coil consisting of a bundle of 80 turns having a radius of 20 cm to produce a magnetic field of 1.5 oersteds?

8. A solenoid of 1 m length consisting of wires wound on a cardboard mailing tube of 10 cm^2 cross-sectional area displays a flux density of 6×10^{-4} weber/meter^2 along its center line. How many turns of wire are required if the current is 3 amperes?

9. A vertical wire carrying current passes through the center of a horizontal drawing board. If the horizontal component of the

earth's magnetic field is .2 oersted, and a small compass needle is used to map out the direction of the magnetic field on the drawing board, a position will be found just east of the wire where the compass will indicate no field intensity, i.e., a neutral point. How far away from the wire is this point if the current in the wire is 10 amperes?

10. The region between the poles of a magnet displays a flux density of .2 weber/meter^2. The pole faces measure 10 cm × 10 cm. A straight wire carrying a current of 12 amperes crossing this region at right angles to the flux is thrust sideways. What is the magnitude of the side thrust?

15

Electromagnetic Induction

The material presented in the preceding chapters was the direct result of discoveries made by early investigators studying the effects of the electric current. What follows is the result of an intuitive idea of Faraday, who inquired about the possible reversibility of the magnetic side-thrust phenomenon, i.e., the possibility of establishing an electromotive force just by the motion of a conductor, and without the aid of a chemical battery.

Induced Electromotive Force. By a series of experiments Faraday discovered quite empirically what would seem to follow logically from today's knowledge of the structure of matter, namely that when a conductor is moved perpendicularly across a magnetic field, the effect is the opposite of side thrust, i.e., a current is induced in the conductor if it is part of a closed electrical circuit. Since the right hand is the opposite of the left hand, and whereas the left-hand rule gives the direction of side thrust when the forefinger points in the direction of the flux and the center finger points in the direction of current; the right hand gives the direction of the induced current when the thumb points in the direction of the motion and the forefinger points in the direction of the flux. See Fig. 89.

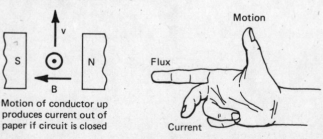

Motion of conductor up
produces current out of
paper if circuit is closed

Fig. 89

The induced electromotive force is given by the following relation:

$$E = BLv$$

from which it follows that the induced current I is given by Ohm's Law $I = E/R$ where R is the resistance of the closed circuit. When B is expressed in weber/meter^2 (mks units), L in meters, and v in meters/sec, E is given in volts. If cgs units are used, E will be expressed in ab-volts, where the ab-volt is 10^{-8} volts, the ab-volt being defined as that electromotive force which is induced when a conductor 1 cm long is moved perpendicularly across a uniform magnetic field of flux density 1 gauss, with a velocity of 1 cm/sec.

Magnetic Flux Viewpoint. It is convenient to picture the preceding situation of a conductor moving across a magnetic field as the conductor "cutting" the magnetic flux lines which represent the field. It can be shown that the electromotive force so induced is proportional to the *time rate* at which the lines are cut, where 1 volt equals 1 weber/sec (mks) and 1 ab-volt equals 1 maxwell/sec (cgs).

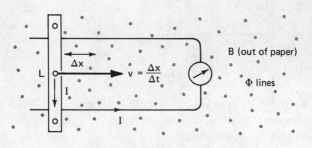

Fig. 90

$$E = BLv = BL \frac{\Delta x}{\Delta t} = B \frac{\Delta A}{\Delta t} = - \frac{\Delta \Phi}{\Delta t}$$

Faraday also showed that, more generally, $\Delta\Phi/\Delta t$ means any change whatever in the number of lines enclosed in a closed loop constituting the circuit, i.e., any manner of flux change, such as an increase or a decrease produced by withdrawing the whole loop from the field, by bringing up a fixed magnet near the loop, by bringing up an electromagnet, by changing the current in a nearby electromagnet, or by changing the permeability of the core of a nearby electromagnet.

Lenz's Law. Lenz's Law states that whenever the magnetic flux associated with an electric circuit is changed by any method whatever, a current is induced in that circuit in such a direction that by its own electromagnetic action it opposes the change that produced it. Thus, induced electromotive force is a back or counterelectromotive force. Hence the negative sign in $E = -\Delta\Phi/\Delta t$.

Mutual Inductance and Self-Inductance. One way to produce a changing magnetic field associated with a circuit is to change the current in a neighboring circuit. The varying magnetic field thus produced in the neighboring circuit causes a varying current to be induced in the first circuit. The effect is referred to as *mutual induction*. See Fig. 91, in which it is evident that by opening and closing the switch K, the varying current thus set

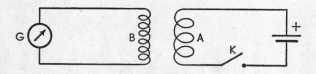

Fig. 91

up in A produces a varying magnetic field which is also common to B, whence a current is induced in B, to be observed by the galvanometer G. The induced electromotive force in B is given by

$$E_B = -\frac{\Delta\Phi}{\Delta t} = -M\frac{\Delta I_A}{\Delta t}$$

where the proportionality factor M is called the *coefficient of mutual inductance*.

The effect is also observed in a single coil due to the mutual effects of all the individual turns of the coil with respect to all the other turns. In this case, the effect is called *self-inductance*, for which the induced electromotive force is expressed by the relation

$$E = - L \frac{\Delta I}{\Delta t}$$

where L is the coefficient of *self-inductance*. This electromotive force, by Lenz's Law, is a back electromotive force which tends to oppose the original change in current. Thus any attempted change in current in a coil which has appreciable self-inductance (due to many turns of wire or due to a core of iron with large permeability) is opposed by the back electromotive force induced. It is as if the electrical property called inductance plays a role in electrical circuits similar to that played by mass (inertia) in a mechanical circuit. Refer back to the concept of mass and the law of Newton in mechanics, recalling that

$$F = ma = m \frac{\Delta v}{\Delta t}$$

where v is the time rate of displacement. In electricity I is the time rate of flow of charge in the expression $E = - L(\Delta I / \Delta t)$.

Carrying the analogy with mechanics further, recall that the kinetic energy is given by the expression W (work) $= \frac{1}{2}mv^2$. In electricity the energy required to set up the magnetic field about a current is given by the expression:

$$W = \frac{1}{2}LI^2$$

When a current-carrying circuit is broken, as by opening a switch, this energy is liberated and often shows up in the form of a visible arc produced at the switch.

Units. The unit of self-inductance, or mutual inductance, is the *henry* in the mks system, or the *ab-henry* in the absolute magnetic system.

The Transformer. The transformer utilizes the principle of induced electromotive force to step up or step down varying voltages, by winding two coils with different numbers of turns on the same core as in Fig. 92. The one coil is called the *primary*, and the other is called the *secondary*. A source of varying electromotive force impressed across the primary coil produces, in the primary circuit, a varying current which in turn produces a varying field that is contained largely within the common core. The turns of the secondary coil loop this varying field and experience an induced electromotive force that is proportional to its number of turns. The relationships are

$$E_s = -N_s \frac{\Delta\Phi}{\Delta t} \qquad \therefore \quad \frac{\Delta\Phi}{\Delta t} = -\frac{E_s}{N_s}$$

$$E_p = -N_p \frac{\Delta\Phi}{\Delta t} \qquad \therefore \quad \frac{\Delta\Phi}{\Delta t} = -\frac{E_p}{N_p}$$

$$\therefore \quad \frac{E_s}{E_p} = \frac{N_s}{N_p}$$

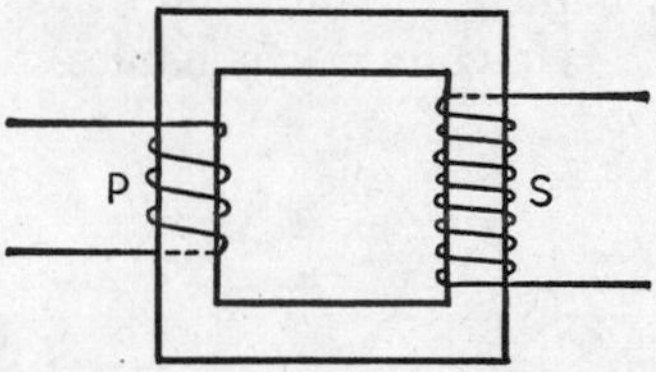

Fig. 92

Thus voltages can be stepped up or down in direct proportion to the number of turns in the secondary as compared to the number of turns in the primary. The transformer as such cannot be used with steady currents such as battery currents. Supplemented by a mechanical breaker system to produce varying currents, the principle of the transformer is utilized to step up ordinary battery currents in the device known as the induction coil.

Problem Procedures

It should be realized that, basically, induced electromotive force is determined merely by calculating the rate at which the magnetic flux is changed. In terms of flux density $B = \Phi/A$ it follows that $\Phi = BA$. Moreover $B = \mu H$.

$$\therefore \quad \Phi = \mu HA \qquad \text{Also} \quad E = -\frac{\Delta\Phi}{\Delta t}$$

In some problem situations, the induced electromotive force is readily determined simply by evaluating the flux passing through the circuit before and then after the change has taken place, and dividing the difference by the time required for the change to take place. Obviously the negative sign need not be considered in numerical problems since only the magnitude of E is involved.

Problem

Consider a plane circular coil consisting of a single turn of radius 20 cm lying flat on a table in the earth's magnetic field whose vertical component is .55 oersted. Turn it completely over so as to leave it upside down in .1 sec. How much electromotive force is induced between the terminals of the coil?

Solution

Note that as the coil lies flat on the table an amount of flux threads through it due to the vertical component of the earth's field according to the following:

$$\Phi = \mu H A$$

where $\mu = 1$ for air, $H = H_v = .55$, and $A = \pi r^2 = \pi (20)^2 = 400\pi$ cm².

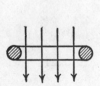

Fig. 93

As the coil is turned over from a horizontal position, through a vertical position to a horizontal position again, it experiences two complete changes of flux. In the vertical position there is no flux threading through it. Thus as the coil is turned through 90°, the change in flux is a change from total to zero flux. In the second 90°, this change is duplicated. Therefore, to turn the coil over, i.e., to rotate it through 180°, the total flux will change twice.

$$\Delta\Phi = 2\mu H A$$

and if Δt is .1 second:

$$E = \frac{\Delta\Phi}{\Delta t} = \frac{2\mu HA}{.1} = \frac{2 \times 1 \times .55(400\pi)}{.1} = 13800 \text{ ab-volts}$$
$$= .000138 \text{ volts} \qquad Ans.$$

Solution in Mks Units

Data: $r = 2 \times 10^{-1}$ meter; $B = .55$ gauss $= .55 \times 10^{-4}$ weber/ meter^2 $= 5.5 \times 10^{-5}$ weber/meter^2; $A = \pi(2 \times 10^{-1})^2 = 4\pi \times 10^{-2}$ meter^2

$$\therefore \quad \Delta\Phi = 2BA \qquad \Delta t = 10^{-1} \text{ sec}$$

$$E = \frac{\Delta\Phi}{\Delta t} = \frac{2 \times 5.5 \times 10^{-5} \times 4\pi \times 10^{-2}}{10^{-1}} = 138 \times 10^{-6} \text{ volt}$$
$$= .000138 \text{ volt} \quad Check.$$

Problem Procedures Continued

A typical problem situation involves the "cutting" of lines of flux by a wire which is part of a closed circuit. When a wire cuts across a field as in Fig. 94, where the rectangle represents

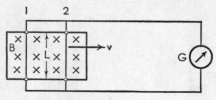

Fig. 94

the south pole of a magnet into which flux lines point, and moves from position (1) to position (2) in reality, the number of flux lines enclosed by the circuit loop is changed. Yet to count the change does not seem so natural as to refer to the velocity with which the wire cuts across the field. In this case the relation $E = -\Delta\Phi/\Delta t$ can be written

$$E = BLv$$

where L is the length of wire perpendicular to and in the field, and v is the velocity of the wire.

Problem

A copper bar laid perpendicularly across a pair of parallel metal tracks running north-south, one meter apart, is slid in a northerly direction. At one end, the tracks are connected through a galvanometer. At the given location the vertical component of the earth's magnetic field is .55 oersted. Is there any reason to suppose that the galvanometer will read a deflection as the bar is moved at the rate of 150 cm/sec? If so, what is it, which way does the current point, how much electromotive force is developed in the bar between the rails, and which end of the bar is at the higher potential?

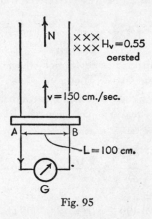

Fig. 95

Solution

Draw diagram. Note that as the bar moves along the track, as indicated, it cuts the vertical component of the earth's field, which points downward, or into the paper. Hence, an electromotive force must be generated between the points A and B, and a current must be established in the circuit, to be detected by the galvanometer. The direction of this current follows from the right-hand rule. Point the forefinger downward, point the thumb northward, and the center finger shows that the current points from B to A in the bar. This means that the point A is at a higher potential than point B. To determine the magnitude of the electromotive force induced, use the relationship

$$E = -BLv$$

where B here is 5.5×10^{-5} weber/meter^2, $L = 1.0$ m, and $v = 1.5$

meter/sec. It follows that, disregarding the negative sign as noted previously

$$E = 5.5 \times 10^{-5} \times 1.5 = 8.25 \times 10^{-5} \text{ volts}$$

Problem

A square coil of 50 cm^2 area consisting of 20 loops is rotated about a transverse axis at the rate of 2 turns per second in a uniform magnetic field of flux density 500 gauss. The coil has a resistance of 20 ohms. What is the average electromotive force induced in the coil during one quarter of a revolution? How much charge is passed through the coil in this interval? What is the average current in a complete cycle?

Solution

Note that in one quarter of a revolution, the flux threading the coil is completely changed once. Therefore $\Phi = NAB$ represents the number of lines cut in $\frac{1}{8}$ second, since $\frac{1}{8}$ second is one quarter of the period of rotation, and since N is the number of loops intercepting the changing flux at all times.

$$\therefore \quad E = \frac{\Delta\Phi}{\Delta t} = \frac{NAB}{\frac{1}{8}} = \frac{20(50)(500)}{\frac{1}{8}}$$
$$= 2 \times 5 \times 5 \times 8 \times 10^4$$
$$= 400 \times 10^4 = 4000000 \text{ ab-volts}$$
$$= .04 \text{ volt} \quad Ans.$$

By Ohm's Law $\quad I = \dfrac{E}{R}$

$$= \frac{.04}{20} = .002 \text{ ampere}$$

But $\quad Q = It = .002(\frac{1}{8}) = .00025 \text{ coulomb} \quad Ans.$

Comment. Note that in this problem the procedure is not so much a matter of substituting values directly into a given formula as it is to reason out how many flux lines are cut per second. Of course, a formula might have been found in a handbook for precisely this situation, but it is better to base the solution upon the basic nature of the concept of induced electromotive force, Ohm's Law, and the basic relation between charge and current.

Problem

A closed loop of wire is placed in a magnetic field in such a way that flux points into the coil as shown in Fig. 96, where the x's indicate the tail ends of flux lines. If the flux density suddenly increases, will current be induced in the loop, and if so will it point clockwise or counterclockwise?

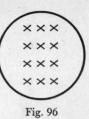

Fig. 96

Solution

This problem is obviously a straight application of Lenz's Law, by virtue of which a current most certainly will be induced in the loop.

The direction of the current is found by noting that a change in flux induces a current which by its own electromagnetic action opposes the change responsible for it. This suggests that the current, whatever its direction, develops a magnetic field out of the paper inside the loop, i.e., opposed to the original field. For this to happen the current must obviously point in a counterclockwise direction.   *Ans.*

Problem

Calculate the self-inductance of a coil if a rate of change of current of 5 amperes per second produces a back emf of 2 volts.

Solution

Since $E = -L \ (\Delta I/\Delta t)$, and again noting that we are dealing with the magnitude only of L:

$$L = \frac{E}{\Delta I/\Delta t} = \frac{2}{5} = .4 \text{ henry} \quad Ans.$$

Problem

The primary of a transformer, consisting of 20 turns, is connected to a varying voltage which a voltmeter indicates to have a value of 110 volts. If the secondary circuit has 1000 turns and a resistance of 20,000 ohms, what voltage would a meter read if connected across it? What is the current in the primary if 100% efficiency is assumed?

Solution

According to the principle of the transformer

$$\frac{E_s}{E_p} = \frac{N_s}{N_p}$$

$$\therefore \quad E_s = \frac{N_s}{N_p} E_p = \frac{1000}{20} (110) = 5500 \text{ volts}$$

If 100% efficiency is assumed, the power output is equal to the power input, where $P = EI$.

By Ohm's Law $\quad I_s = \dfrac{E_s}{R_s} = \dfrac{5500}{20000} = .275$ ampere

$$\therefore \quad P = E_s I_s = E_p I_p$$

$$\therefore \quad I_p = \frac{E_s I_s}{E_p} = \frac{5500}{110} (.275)$$

$$= 50(.275) = 13.75 \text{ amperes} \qquad Ans.$$

Problems: Induced Electromotive Force

1. What is the average emf in volts developed in a circular coil of 5×10^{-2} meter radius and with 40 turns, when it is rotated from a position normal to a field of 3×10^{-5} weber/meter^2 in air to a position parallel to that field in .25 sec?

2. What is the mutual inductance between two adjacent coils if an emf of 125 volts is induced in one when the current in the other is changing at the rate of 25 ampere/sec?

3. A circular coil of 25 turns and 10^{-2} meter in diameter is located between the pole pieces of a magnet and normal to the magnetic field of 3.18 weber/meter^2. If the coil is removed in .2 sec, what is the induced emf?

4. A square coil, 5 cm along each side, wound with 50 turns, is located between the pole pieces of a magnet and perpendicular to the uniform magnetic field of 20,000 oersteds. If the coil is removed in .5 sec, what is the induced emf?

5. What is the induced emf in a given coil when the current in a neighboring coil is changing at the rate of .5 ampere/sec because of the closing of a switch connecting it to a 2 volt cell? The mutual inductance between the coils is 2 henries.

6. Find the emf induced in a car axle of 1.3 meter length when the car is running north in the magnetic meridian with a speed of

4×10^{-1} meter/sec. The vertical component of the earth's field is $.55 \times 10^{-4}$ weber/meter². Express the answer in the proper units.

7. A coil composed of 20 turns of wire lies on a table top. As the north pole of a magnet is brought downward toward it, the flux changes at the rate of 500,000 lines per sec. What is the magnitude of the induced emf in the coil?

8. A wire 100 cm long lying perpendicular to a magnetic field of 100 gauss is moved perpendicularly across the field at a speed of 100 cm/sec. What is the magnitude of the induced emf?

9. With what speed must a wire 5 cm long be moved perpendicular to a magnetic field of 1000 gauss to produce an emf of .01 volt?

10. What is the final steady current value when the switch connecting a 2 volt cell to a coil of 50 ohms resistance and self-inductance of 25 henries is closed?

11. A coil whose face area is 25 cm² is jerked out from between the jaws of a permanent magnet. If the process requires .2 sec, and an electromotive force of 10^{-4} volts is induced in the coil, what is the flux density between the jaws of the magnet?

12. A circular coil of 50 turns having a radius of 20 cm lies on a horizontal table top. It is turned over in .25 sec at a place where the earth's dip angle is 72° and where the horizontal component of the earth's field intensity is .20 oersted. If the coil has a resistance of 100 ohms, what average current is induced in the coil?

13. A square loop 10 cm on a side is rotated about an axis passing through the center and parallel to a side, between the poles of an electromagnet having a flux density of 15,000 gauss. If it is rotated at a speed of one complete turn per second, what is the maximum electromotive force generated in it?

14. An iron core solenoid has 80 millihenries of self-inductance. How much back electromotive force is induced in this coil if a current of 2 amperes requires .4 sec to be established as a switch is closed connecting it to a battery?

15. A wire which closes a circuit with a galvanometer is passed perpendicularly across a magnetic field of intensity 15,000 oersteds with a speed of 100 cm/sec. If only 10 cm of the wire are exposed to the field, and if the circuit has a resistance of 200 ohms, how much current is induced in it?

16. A primary coil and a secondary coil are independently wound on the same iron core. The former has 100 turns and the latter has 500 turns. A battery connected to the primary establishes a current of 2 amperes in .1 sec, resulting in an electromotive

force of 550 volts being induced in the secondary. (a) What is the mutual inductance of the circuits? (b) What counterelectromotive force is induced in the primary?

17. A solenoid has a self-inductance of .5 henry and a resistance of 50 ohms. Approximately how long does it take for the current to build up from zero to 2 amperes when the solenoid is connected to a battery, if 2 amperes is about half the steady value the current ultimately becomes?

18. The flux through a single turn of wire changes from one thousand to one million maxwells in .1 sec. What electromotive force in volts is induced in the wire?

Alternating Current. When a loop of wire is rotated continuously about a transverse axis across a constant magnetic field, a sinusoidally varying electromotive force is induced in the coil, producing a sinusoidally varying current called "alternating" current. This is the type of current commonly called ac. Also the so-called 110 volt ac is such an "alternating" electromotive force.

Although the average value of an alternating current, taken over any number of complete cycles, is necessarily zero, it does not follow that the effective value is also zero. Since heating, for example, depends upon the second power of the current, and since the square of the negative components of alternating current is always positive, it is evident that heating by such a current does not average out to zero. The effective value of such a current is defined as the square root of the average squared value of the current. For sinusoidally varying currents the *effective value*, sometimes called RMS (root-mean-square) value, is given by $I_{eff} = .707 I_{max}$. Similarly $E_{eff} = .707 E_{max}$. Ac ammeters and ac voltmeters read such effective values. Thus 110 volt ac refers to a voltage which varies from 0 to 156 volts to 0 to −156 volts to 0, etc.

Ac Impedance. Since ac current is a varying current, back electromotive force is always present depending upon the amount of self-inductance L in the circuit. Hence inductance as well as resistance enters into Ohm's Law considerations for ac circuits. Moreover, capacitance also is involved, because alternating currents oscillate back and forth around condensers which, in dc circuits, constitute breaks in the line. Whereas in dc circuits $I = E/R$, in ac circuits $I = E/Z$, where Z is called *impedance*, which is related to resistance, inductance, and capacitance in the following manner:

$$Z = \sqrt{R^2 + X^2} \quad \text{where } X \text{ is called reactance}$$

There are two kinds of reactance: X_L (inductive reactance) = $2\pi nL$, and X_C (capacitive reactance) = $1/2\pi nC$.

Thus
$$Z = \sqrt{R^2 + \left(2\pi nL - \frac{1}{2\pi nC}\right)^2}$$

Impedance, reactance, and resistance are all measured in ohms.

Lagging and Leading Currents. A preponderance of inductive reactance in a series circuit causes the current to lag behind the electromotive force; a preponderance of capacitive reactance causes the current to lead. The absence of reactance, as in a purely resistive circuit, results in a current which follows the electromotive force and remains in phase with it. See Fig. 97.

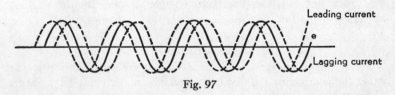

Fig. 97

It is to be noted that the effect of capacitive reactance tends to offset the effect of inductive reactance:

$$X = 2\pi nL - \frac{1}{2\pi nC}$$

There is a particular frequency, for each value of L and C, for which the capacitive reactance exactly balances the inductive reactance. This frequency is given by $n = (1/2\pi)\sqrt{1/LC}$. Corresponding to this frequency a series circuit behaves like a purely resistive circuit, and the current acquires its maximum effectiveness. This frequency is called the *resonant frequency*. Tuning a high frequency radio circuit is a matter of adjusting L and C for resonance.

Ac Power. Because of the effects of lag and lead, the average value of power consumption in an ac circuit is not a simple product of the voltage and the current, as it is in the case of the dc circuit. It is given by the expression

$$\bar{p} = E_v I_a \cos \phi$$

where $\bar{p}$ stands for average power, E_v represents the voltmeter reading, I_a is the ammeter reading, and ϕ is the angle of lag or lead as the case may be. The factor $\cos \phi$ is called the *power factor*.

Problem Procedures

It should be noted in considering alternating current problems that a capacitance does not constitute an open circuit as it does in dc circuitry. Also, in dc work, the resistance can be determined by the ratio E/I, whereas this ratio in ac work gives the impedance, which is more than resistance alone, unless the capacitive reactance just balances the inductive reactance. Incidentally, ac meters will read on dc, but dc meters will not read on ac.

Problem

A coil having resistance and inductance is connected in series with an ac ammeter across a 100 volt dc line. The meter reads 1.1 amperes. The combination is then connected across a 110 volt ac 60 cycle line and the meter reads .55 ampere. What are the resistance, the impedance, the reactance, and the inductance of the coil?

Solution

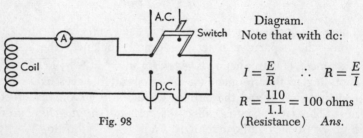

Fig. 98

Diagram.
Note that with dc:

$$I = \frac{E}{R} \quad \therefore \quad R = \frac{E}{I}$$

$$R = \frac{110}{1.1} = 100 \text{ ohms}$$

(Resistance) *Ans.*

On ac, however: $\dfrac{E}{I} = Z = \dfrac{110}{.55} = 200$ ohms (Impedance)   *Ans.*

But $\qquad Z = \sqrt{R^2 + X^2}$ $\qquad \therefore \quad 200^2 = 100^2 + X^2$

$\therefore \quad X^2 = 40000 - 10000 = 30000$

$\qquad\qquad X = \sqrt{30000} = 173$ ohms (Reactance) $\quad$ Ans.

And $\qquad X = 2\pi nL$

$\therefore \quad L = \dfrac{X}{2\pi n} = \dfrac{173}{2\pi(60)} = .459$ henry (Inductance) $\quad$ Ans.

Problem

A radio receiving set is tuned to a certain station by the use of a .25 millihenry (2.5×10^{-4} henry) inductance, and a 32.2 micro-microfarad (32.2×10^{-12} farad) condenser. What is the frequency of the station? What is its wave length?

Solution

For a tuned circuit, the resonant frequency is given by

$$n = \frac{1}{2\pi} \sqrt{\frac{1}{LC}}$$

$$\therefore \quad n = \frac{1}{2\pi} \sqrt{\frac{1}{2.5 \times 10^{-4} \times 32.2 \times 10^{-12}}} = \frac{1}{2\pi} \sqrt{\frac{10^{16}}{81}}$$

$$n = \frac{10^8}{2\pi 9} = \frac{10^8}{56.5} = 1.77 \times 10^6 \text{ cycles per sec} \quad Ans.$$

The velocity of radio waves is 3×10^{10} cm/sec

and $\qquad v = n\lambda$ for all waves

$$\therefore \quad \lambda = \frac{v}{n} = \frac{3 \times 10^{10}}{1.77 \times 10^6} = 1.7 \times 10^4 = 17,000 \text{ cm}$$

$$= 170 \text{ m} \quad Ans.$$

Problem

What is the power factor of a circuit if the following meter readings prevail? Ammeter reads .15 amperes, voltmeter reads 115 volts, and wattmeter reads 15.9 watts.

Solution

Power factor ($\cos \phi$) is related to power by the expression

$$\bar{p} = E_v I_a \cos \phi$$

$$\therefore \quad \cos \phi = \frac{\bar{p}}{EI} = \frac{15.9}{115(.15)} = .92 \quad Ans.$$

Comment. This corresponds to a phase lag angle of current behind voltage of 23° and must be due to inductive reactance.

Problems: Alternating Current

1. When an ac ammeter in a circuit reads 6 amperes on 60 cycle ac, what is (a) the RMS value of the current? (b) the maximum (peak) value of the current? (c) the average value of the current?
2. If an impedance of 345 ohms is connected to 115 volts 60 cycle ac, what is the resulting current through the impedance?
3. What is the impedance of a coil which has a resistance of 30 ohms and a reactance of 40 ohms on 115 volt 60 cycle ac?
4. A hot plate is marked 115 volts, 60 cycle, 10 amperes. (a) What is the RMS value of the current through it when connected to 115 volts 60 cycle? (b) What is the peak current through it?
5. An appliance which is marked 115 volts, 60 cycle ac, 750 watts is found upon test to draw 9 amperes. What is its power factor?
6. A transformer is designed to deliver .200 ampere at 600 volts when the primary is connected to 115 volts 60 cycles. What current will the primary draw from the line under full load? Assume 100% efficiency.
7. The primary coil of an ideal transformer has 6000 turns of wire and the secondary coil has 600 turns. A potential difference of 550 volts is established across the ends of the primary coil. What current will be in the secondary circuit if its impedance is 27.5 ohms?
8. A coil draws 10 amperes on a 110 volt 60 cycle line. When this same coil is connected in series with a 4 ohm resistance on a 110 volt dc line, the current drawn is 15 amperes. Calculate (a) the impedance, (b) the resistances, and (c) the inductance of the coil.
9. The current in the secondary of a 20 to 1 transformer is 40 amperes while the voltage across the primary windings is 2200 volts. What is (a) the current in the primary? (b) the voltage across the secondary? (c) the power output? Assume 100% efficiency of the transformer.
10. What is the resonant frequency of a circuit consisting of a capacitance of 50 micro-microfarads and an inductance of 60 milli-henries?
11. An appliance rated at 115 watts is plugged into a 115 volt line and is found to draw 1.15 amperes. What is the power factor?
12. What is the impedance of a choke coil having an inductance of .1 henry on 60 cycles and a resistance of 40 ohms?

Part V.

Light

16

The Nature of Light

Although the subject of light has been studied extensively for centuries and many of its aspects were catalogued by early scientists, only in the present century has the real nature of light been discovered. Experimental evidence has supported at various times a corpuscular theory, and at other times a wave theory of its nature. As recently as the early 1920s the difficulties of reconciling these two opposing theories seemed insurmountable, but soon thereafter the duality was satisfactorily resolved by the concept of the *photon*, treated as a bundle of waves.

It is not the purpose of this text to dwell on the question of the nature of light. The reason for introducing the question is rather to point up the fact that a strictly logical approach to the study of light is not so simple as the approach to the study of electricity. There the subject developed logically once the concept of electricity was established.

Geometric vs. Physical Optics. The study of light divides naturally into two broad categories. On the one hand is that part known as geometric optics, which deals with the straight-line propagation of light and the laws of regular reflection and regular refraction. These laws, which involve straight lines and angles (geometry), explain the formation of optical images by optical instruments. On the other hand, physical optics deals with those aspects of light which are more closely related to the nature of light, such as interference, diffraction, and color. These terms will be considered later. In this text more attention will be devoted to geometric optics than to physical optics, primarily because the former lends itself more readily to numerical problems than does the latter.

Photometry. It is logical to start with the concept of light traveling in straight lines in all directions from a pointlike source. To express quantitatively the amount of light emitted by such a source, the idea of enveloping the source by a spherical surface to collect all the light falling on it is suggested. It then seems natural to consider as a unit amount of light that amount which is emitted in a unit solid angle (*ster-radian*) by a standard source (*candle*). Such a unit of light flux is known as a *lumen*.

The ster-radian is in solid geometry what a radian of angular measure is in plane geometry. Whereas a radian is an angle equal to $1/2\pi$th of a circle, a ster-radian is a solid angle, a cone equal to $1/4\pi$th of a sphere, i.e., it is a cone of such size that 4π of them can be clustered about a point in space.

Intensity of a Source. The *candle*, until recently, has referred to the *intensity* of a prescribed candlelike flame from a standardized lamp, whereupon the lumen is the amount of light emitted by the standard candle in a ster-radian. Obviously the lumen is a unit of radiant power which is also expressible in watts. Actually 685 lumens of green light correspond to one watt (an earlier figure was 621). Today the lumen is defined in terms of $\frac{1}{685}$ watts, for green light, and the candle is that intensity of pointlike source capable of emitting one lumen in one ster-radian. Thus the number of lumens F emitted by a point source of intensity I is $F = 4\pi I$.

Luminance or Brightness. Whereas a point source has intensity, an extended source has *luminance* (formerly called *brightness*). It is expressed in candles per unit area. Another unit of luminance is the *lambert*, which is used for matte or diffusely reflecting surfaces characterized by the fact that the luminance does not depend upon the angle at which it is viewed.

Illuminance or Illumination. When light is received by a surface, the surface is said to have illuminance, or be illuminated. The illuminance E is the number of lumens received per unit area. At a distance of one foot from a candle source, the illuminance is one lumen per square foot, or one foot-candle. It can be shown in general that $E = (I/r^2) \cos \theta$, where I is the intensity in candles, r is the distance from source to surface, and θ is the angle of incidence of the light on the receiving surface, i.e., the angle between the ray and the normal to the surface.

Photometric Measurements. The intensity of a point source may be compared to that of a standard source by determining the distance it must be placed from a screen to produce the same illuminance that the standard source produces at a known distance. Applying the inverse square law for normal incidence

$$E = \frac{I_1}{r_1{}^2} = \frac{I_2}{r_2{}^2}$$

where the subscripts 1 and 2 refer to the standard and the unknown source respectively.

$$\therefore \quad \frac{I_1}{I_2} = \frac{r_1{}^2}{r_2{}^2}$$

Problem
How many lumens of light are emitted by a 10 candle source?

Solution
On the basis of the above definitions, there must be one lumen per ster-radian from one candle. Moreover, the space about a point comprises 4π ster-radians, or $F = 4\pi I$.

$$\therefore \quad F = 4\pi(10) = 125.6 \text{ lumens} \qquad Ans.$$

Problem
What is the illuminance 10 ft directly beneath a 10 candle source?

Solution
This is a straightforward application of the inverse square law of illuminance.

$$E = \frac{I}{r^2}$$

$$\therefore \quad E = \frac{10}{10^2} = \frac{1}{10} \text{ lumen/sq ft} \qquad or \qquad \tfrac{1}{10} \text{ foot-candle} \qquad Ans.$$

Problem
What is the illuminance on a table at a distance of 5 ft from a point 10 ft directly beneath the 10 candle source in the previous problem?

Solution

In this case the distance from the source to the point in question is

$$\sqrt{10^2 + 5^2} = \sqrt{125} = 11.2 \text{ ft (approx)}$$

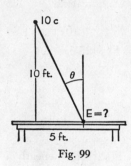

Fig. 99

Moreover, the light is incident at an angle θ, whose tangent is $\frac{5}{10} = .50$

$$\therefore \quad \theta = 27° \text{ (approx)} \quad \text{and}$$
$$\cos \theta = .89 \text{ (approx)}$$

In general, $\qquad E = \dfrac{I}{r^2} \cos \theta.$

$$\therefore \quad E = \frac{10}{125} (.89) = .071 \text{ foot-candle} \qquad \textit{Ans.}$$

Problem

At what distance from a screen must a 10 candle lamp be placed to produce the same illuminance that will be produced by a 15 candle lamp held 40 cm from the same screen?

Solution

Since $\qquad \dfrac{I_1}{I_2} = \dfrac{r_1{}^2}{r_2{}^2}$

it follows that

$$r_2{}^2 = \frac{I_2}{I_1} r_1{}^2$$

$I_1 = 15$ candles, $I_2 = 10$ candles, and $r_1 = 40$ cm.

$$\therefore \quad r_2{}^2 = \tfrac{10}{15}(40)^2 = 1067$$
$$r^2 = \sqrt{1067} = 32.7 \text{ cm (approx)} \qquad \textit{Ans.}$$

Problem

If an illumination of 10 foot-candles is desired for reading purposes, how far should one hold the printed page from a 100 watt lamp which has an efficiency of 1.25 candles per watt?

Solution

At 1.25 candles per watt, the 100 watt lamp is equivalent to a 125 candle source.

Since $E = I/r^2$ where $E = 10$ foot-candles, and $I = 125$ candles

$$r^2 = \frac{I}{E} = \frac{125}{10} = 12.5 \qquad \text{or} \qquad r = \sqrt{12.5} = 3.5^+ \text{ ft} \qquad Ans.$$

Problems: Light and Photometry

1. What is the illumination in foot-candles 5 ft directly beneath a 100 candle source?

2. The illumination on a table top 3 ft directly below a point source is 10 lumens/sq ft. If the source be moved 2 ft higher, what will be the illumination on the table top?

3. A standard lamp of 20 candles is located 2 m from a comparison lamp. A balance point is found 70 cm from the standard. An unknown lamp is substituted for the standard and a balance is obtained 90 cm from the unknown lamp. What is the candle rating of the unknown bulb?

4. A standard lamp of 40 candle power is located 210 cm from a comparison lamp. An illumination balance is obtained on a photometer screen when it is placed between them and 70 cm from the comparison lamp. What is the candle power of the comparison lamp?

5. A 60 watt lamp has a luminous efficiency of 10 lumens per watt. (a) What is the value of the flux emitted by the lamp? (b) What is the candle power of the lamp, assuming the light is emitted uniformly in all directions? (c) If the illumination on a screen due to this lamp is 4 foot-candles, what is the illumination on the screen when the distance from screen to lamp is doubled?

6. A lamp of unknown candle power is placed 100 cm from a lamp of 40 candle power. The illumination is the same on both sides of a photometer screen when the screen is 40 cm from the 40 candle

power lamp. (a) What is the candle power of the unknown lamp? (b) What is the value of the illumination on either side of the screen?

7. The illumination of bright sunlight is 9,600 lumens/ft². What is the total flux incident on a pane of glass 2 ft by 3 ft when the plane of the glass is perpendicular to the rays from the sun?

8. The total light flux from a certain carbon arc is calculated to be 12,000 lumens. Assuming the arc to be a point source, what must be its intensity?

9. The standard light source of a photometer has a candle power of 50 candles. What is the illumination on a screen 75 cm away?

10. A surface is illuminated at a level of 4 foot-candles by a 36 candle power lamp. How far away must it be?

11. What is the illuminance (illumination) produced by a spotlight on a floor if the area of the spot is 2 sq ft and if the beam is produced by placing a reflector behind a pointlike source of 10 candle power intensity so as to collect all the emitted light into a parallel beam? Assume no absorption of light by the air between lamp and floor.

12. A 20 candle power lamp and a lamp of unknown intensity are mounted on a photometer bench 200 cm apart. What is the candle power intensity of the unknown lamp if a screen placed 70 cm from the known lamp is uniformly illuminated?

13. A light source consists of a cluster of lamps made up of two 60 watt lamps and one 100 watt lamp placed 6 ft from a surface. What is the illuminance at the surface, assuming an efficiency of 10 lumens per watt for each light?

14. If a pointlike source of light produces an illumination of 5 foot-candles at a distance of 4 ft, what illumination results if the source is moved 4 ft farther away?

15. A spotlight projects a beam of 16 lumens per sq ft on a table in such a way that the beam makes an angle of 60° with the horizontal. What is the illuminance where the beam strikes the table?

16. A 32 candle power lamp is 4 ft directly above a desk. A second lamp is 4 ft above the desk, but 10 ft horizontally away from the second lamp. What is the intensity of the second lamp, if the two together produce twice the illuminance of one alone at the desk?

17. A focusing flashlight concentrates 50 lumens per sq ft on a surface 6 ft away. If the reflector and the lens of the flashlight were to be removed, what candle power lamp would be required to produce the same illumination as before?

18. A 32 candle power automobile lamp is placed at the center of

curvature of a concave mirror which subtends exactly 2 ster-radians of solid angle. How much flux does the mirror receive?

Law of Regular Reflection. When a light ray strikes a mirror-like surface, it is reflected in such a manner that the angle of reflection, measured with respect to the normal (perpendicular) to the surface, is equal to the angle of incidence also measured with respect to the normal. Moreover, both angles and the normal lie in the same plane. This phenomenon is known as *regular reflection,* and this law of angles is known as the law of regular reflection. See Fig. 100.

$$r = i$$

Law of Refraction. When a light ray is incident upon the boundary surface between two transparent media, part of it is reflected according to the law of regular reflection, but part is refracted, i.e., transmitted into the second medium. Depending upon whether the second medium is optically more or less dense than the first medium, the velocity of the light is decreased or increased as it crosses the boundary, with the result that if the ray is oblique, it experiences a change in direction. The ratio of the sine of the angle of incidence to the sine of the angle of refraction (both angles measured with respect to the normal) was found by Snell to be constant. The constant is known as the refractive index of the second medium with respect to the first medium, although on the wave theory the absolute refractive index of a medium is the ratio of the velocity of light in vacuum to the velocity of light in the given medium. See Fig. 101.

$$n = \frac{\sin i}{\sin r} \text{ (Snell's Law)}$$

It is better to write $n = n_2/n_1$, where n_2 is the absolute refractive index of the second medium and n_1 is the refractive index of the first medium. A more general form of Snell's Law then follows:

$$n_1 \sin i = n_2 \sin r$$

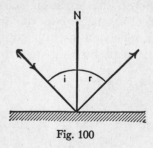

Fig. 100

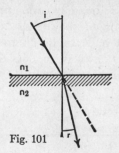

Fig. 101

It is to be noted that when a ray of light passes from a medium of low index (optically rarer medium) to a medium of higher index (optically denser medium), it is bent toward the normal; and vice versa, a ray passing from a denser to a rarer medium is bent away from the normal.

Total Internal Reflection. Considering further the case of light passing from a denser into a rarer medium (from a medium of greater to a medium of lesser refractive index), it is evident that the angle of refraction cannot exceed 90°, for which angle there corresponds a certain angle of incidence, i_c, given by Snell's Law. See Fig. 102.

$$n_1 \sin i = n_2 \sin r$$

$$\sin i = \frac{n_2}{n_1} \sin r$$

$$\therefore \quad \sin i_c = \frac{n_2}{n_1} \text{ (where } r = 90°\text{)}$$

Angles of incidence greater than i_c would require angles of refraction greater than 90°, which are impossible. Thus i_c is a critical angle of incidence.

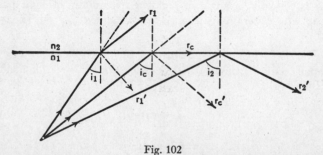

Fig. 102

It is found experimentally that when the angle of incidence exceeds i_c, the result is total reflection back into the initial medium. Actually there is partial reflection corresponding to angles of incidence less than i_c, but for all angles greater than i_c the reflection is complete.

In the case of glass–air:

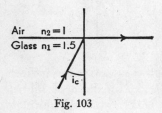

Fig. 103

$$\sin i_c = \frac{n_2}{n_1} = \frac{1}{1.5} = .667$$
$$\therefore \quad i_c = 42° \text{ (approx)}$$

Hence total internal reflection occurs for all incident angles greater than $42°$.

Problem

How fast does light travel in glass of refractive index 1.5?

Solution

By definition, refractive index n is the ratio of the velocity of light in vacuum $(3.00 \times 10^{10} \text{ cm/sec})$ to the velocity of light in the medium in question.

$$\therefore \quad n = \frac{3.00 \times 10^{10}}{v} = 1.5$$
$$\therefore \quad v = \frac{3.00 \times 10^{10}}{1.5} = \frac{2}{3}(3.00 \times 10^{10})$$
$$= 2.00 \times 10^{10} \text{ cm/sec} \quad \textit{Ans.}$$

Problem

What is the critical angle of incidence for a ray of light passing from glass into water? Assume $n_{glass} = 1.50$ and $n_{water} = 1.33$.

Solution

By Snell's Law $n_1 \sin i = n_2 \sin r$. Also critical angle of incidence means $\sin r = 1$.

$$\therefore \quad \sin i = \frac{n_2}{n_1} = \frac{1.33}{1.50} = .887$$
$$\therefore \quad i_c = 62° \text{ (approx)} \quad \textit{Ans.}$$

Problems: Regular Reflection and Refraction

1. A beam of light passes from water of refractive index $\frac{4}{3}$ into glass of refractive index $\frac{3}{2}$, making an angle of refraction of 30° with the normal to the surface. What must be the sine of the angle of incidence? (Express in fraction form.)

2. If the angle of incidence of a ray of light traveling from glass ($n = 1.50$) to water ($n = 1.33$) is 45°, what is the sine of the angle of refraction?

3. The velocity of light in vacuum is 3×10^{10} cm/sec. The velocity in a certain type of glass is 1.80×10^{10} cm/sec. What is the index of refraction of this glass?

4. If the critical angle for a ray of light entering air from glass is arc sin .67, what is the index of refraction for the glass?

5. The velocity of light is 3×10^{10} cm/sec. What is the frequency of light waves if the wave length is 555×10^{-7} cm?

6. A layer of water ($n = 1.33$) rests on a layer of chloroform ($n = 1.45$). A ray of light in the water strikes the surface between the liquids at an angle of 30° with the normal. What is the sine of the angle of refraction?

7. A person 6 ft tall stands 4 ft in front of a plane mirror. (a) Where is his image? (b) How tall is his image? (c) What is the shortest mirror in which this person can see his whole image? (d) Does the answer to (c) depend upon how far the person is from the mirror?

8. Through what angle must a plane mirror be rotated in order to rotate an image through 90°?

9. What is the velocity of light in diamond of refractive index 2.4?

10. If the critical angle of incidence for rock salt is 41°, what is its index of refraction?

11. A flash lamp immersed beneath the surface of a liquid displays a cone of light which is intercepted by the surface to form a circle of light. Taking 1.33 as the refractive index of water, what depth is required to form a circle 120 cm in diameter?

12. What is the critical angle for a piece of glass in water if the refractive index of glass is 1.50 and that of water is 1.33?

17

Geometric Optics

When rays of light are incident upon spherical surfaces, they experience changes in direction by refraction and/or reflection to produce interesting and sometimes unusual results. Light rays from an illuminated object placed in front of a spherical surface can be so bent as to make them appear to come from an image of the object. In other words, optical images can be formed by the reflection and refraction of light. By a knowledge of the laws of reflection and refraction, one can design optical instruments to produce images at will and other instruments to serve as visual aids in viewing them. Images of objects can be located by either mathematical or graphical methods, both of which will now be considered as a study in geometric optics. We shall start with a consideration of refraction and reflection at a single spherical surface.

Single Spherical Surface. Consider a point P on the axis of a spherical surface of radius R, and at a distance u from it, as in Fig. 104. It is necessary to adopt a set of conventions in order

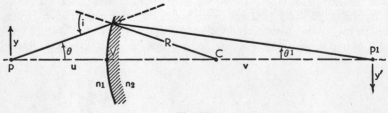

Fig. 104

that algebraic signs may have physical significance. Different conventions are to be found in different textbooks. Any one of

these sets of conventions may be used, but it is absolutely necessary to follow through consistently, whichever one is used. Sign
conventions cannot be changed in the middle of a problem.

Recommended Sign Conventions.

1. Always draw diagrams with light passing from left to right,
i.e., always place object at left of the spherical surface concerned.

2. Assume object distance u (measured to the vertex V) to be
positive when object lies to left of vertex.

3. The image distance v (also measured from V) is positive
when image lies to the right of V, and negative when it lies to the
left of V.

4. Consider the radius of curvature R to be positive when the
center C of the surface lies to the right of the vertex, and negative
when the center lies to the left of the vertex.

5. Measure all angles with respect to the axis of the surface or
a normal to the surface as the case may require. Angles are
positive when they are measured in a counterclockwise manner
from the axis or the normal to the surface. When they are measured in a clockwise manner they are negative.

6. Consider the first medium (the medium to the left of the
surface) to have refractive index n_1 and the second medium (that
to the right) to have refractive index n_2 as in the diagram (see
Fig. 104).

7. Indicate the transverse dimension of the object by y and
consider it positive when it points upward. Let the image size
(transverse dimension) be represented by y^1, which also is
positive if it points upward and negative if it points downward.

Single Spherical Surface: Paraxial-Ray Formula. Referring
back to the diagram in which a point P on the axis of a spherical
surface is imaged at a point P^1 at a distance v from the vertex of
the surface, it can be shown that the following formulas hold

$$\frac{n_1}{u} + \frac{n_2}{v} = \frac{n_2 - n_1}{R}$$

and

$$\frac{y^1}{y} = -\frac{n_1 v}{n_2 u}$$

These are general formulas which hold for very small angles θ
and θ^1 (for so-called paraxial rays). The lateral magnification m

is defined as y^1/y (image size divided by object size).

It will be noted that $m = 1$ means the image has the same size as the object. Diminution does not mean negative magnification, but fractional magnification. Negative magnification merely means that the image is inverted with respect to the object.

Special Cases.

1. The above formulas hold not only for spherical surfaces, but for plane surfaces as well, for which $R = \infty$ and $1/R = 0$.

2. The above formulas hold for reflection whereby light is reflected back into the original medium but reversed in direction. For reflection $n_2 = -n_1$.

Thus for a spherical mirror

$$\frac{n_1}{u} + \frac{(-n_1)}{v} = \frac{-n_1 - n_1}{R}$$

or

$$\frac{1}{u} - \frac{1}{v} = -\frac{2}{R}$$

and

$$m = +\frac{v}{u}$$

Reflection from Spherical Surfaces. Reflection can be studied by the graphical method, sometimes called the ray-diagram method. We shall consider separately the convex mirror, the concave mirror, and the plane mirror.

CONVEX MIRROR. Consider an object O in the form of a vertical arrow in front of, and with its base on the axis of, a convex mirror of radius R, as in Fig. 105. Every point of the object O can be considered to emit rays of light in all directions. We shall

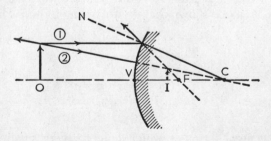

Fig. 105

concentrate on only two such rays emitted by the head end of the arrow: one parallel to the axis and the other pointing toward the center C. The first one is unique in that at reflection, as the angle of reflection equals the angle of incidence, the reflected ray appears to emanate from the point F, called the *focal point*. The second one is unique in that it makes zero angle of incidence with the normal and hence is reflected directly back upon itself, appearing to have come from C. After reflection, these two rays, which actually originated at a common point, appear to have originated at another common point, which is therefore the image of the original point. Thus I is the image of O, because every point of the object is reproduced at the image.

The specific procedure for locating images is as follows.

1. Draw a diagram, with object at left of surface in order that light may proceed from left to right in accordance with the conventions previously listed.

2. Consider two rays from the head of the object, the one parallel to the axis, and the other directed toward the center of curvature of the surface.

3. In the case of the *convex* surface, the parallel ray will be reflected *from* the focal point (halfway between the center and the vertex). In the case of the *concave* surface (see diagram which follows presently) the reflected ray will pass *through* the focal point. Extend it indefinitely.

4. The ray directed toward the center will be reflected straight back upon itself. Extend it indefinitely.

5. The point where the two rays thus reflected intersect is the image of the head of the object. Connect this point with the axis.

CONCAVE MIRROR. Follow the above procedure to draw Fig. 106. In the figure as drawn, the object is so far away that the image is formed at the left and is inverted. Here the image is a *real image* because the light rays actually intersect in space. Such

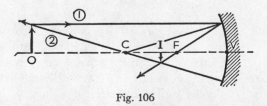

Fig. 106

an image can be caught on a screen placed where it is formed. In the case of the convex mirror above, the intersection was only apparent, not real, because it was behind the mirror where no light is present. Such an image is called a *virtual image*.

If, in the case of the *concave* mirror, the object is relatively close to the mirror, i.e., between the focal point and the vertex as in Fig. 107, the image is also virtual, i.e., to the right of the

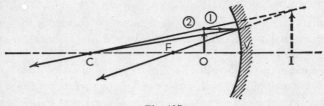

Fig. 107

vertex, and also erect. All *virtual images* from a single surface are *erect*, and all *real images* from a single surface are *inverted*. Thus concave mirrors can form both real (inverted) images and virtual (erect) images depending upon whether the object lies outside the focal point or inside the focal point, whereas convex mirrors can form virtual images only.

PLANE MIRROR. In the case of the plane mirror (of infinite radius) the image is always virtual (erect), and just as far to the right of the mirror as the object is to the left of it. See Fig. 108.

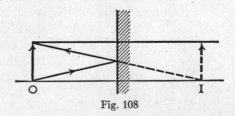

Fig. 108

Incidentally, the images formed by convex mirrors are not only virtual; they are also always reduced in size compared with the object. On the other hand, the virtual images produced by concave mirrors are always enlarged. The real images formed by concave mirrors may be either reduced or enlarged depending upon the object distance compared with the focal length. The virtual images formed by plane mirrors are the same size as the object.

Problem Procedures

Problems involving spherical mirrors invariably have to do with the location of images and the determination of their nature and size as compared with the corresponding objects. It is usually expected that solutions will be obtained by calculations. Such calculations involve the formulas:

$$\frac{1}{u} - \frac{1}{v} = -\frac{2}{R}$$

and

$$m = \frac{v}{u} = \frac{y^1}{y}$$

In all cases, however, a ray diagram should accompany the calculations, even if it is no more than a sketch drawn approximately to scale. It is very easy to confuse algebraic signs, so that the ray diagram serves as a useful check on the algebra as well as to give a certain amount of physical "feel" to the problem situation.

Problem

An object lies on the axis and 20 cm in front of a spherically concave mirror of 10 cm focal length. Locate the image. Is it real or virtual? How large is it compared with the object?

Solution

The word concave when applied to a mirror suggests a mirror which, by the conventions followed in this text, is one whose center of curvature lies to the left of the surface and whose radius of curvature is negative.

RAY-DIAGRAM METHOD

The focal length of a mirror is always taken as one-half the radius of curvature. Consequently, a ray diagram of this problem appears as in Fig. 109.

$F = 10$
$R = -20$

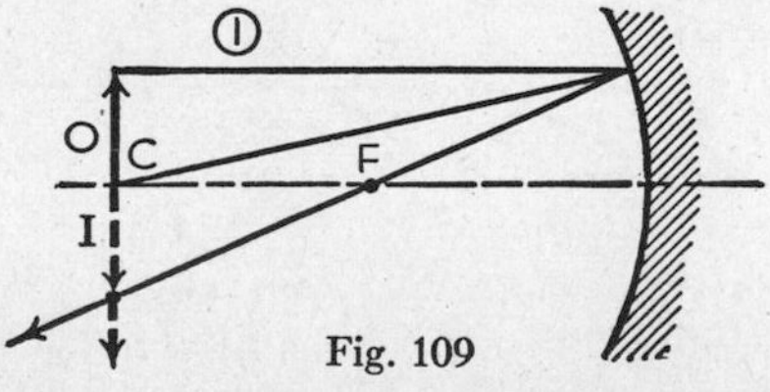

Fig. 109

The image appears to be inverted and located at the same distance from the mirror as the object itself.

FORMULA METHOD

$$\frac{1}{u} - \frac{1}{v} = -\frac{2}{R}$$

$$\frac{1}{20} - \frac{1}{v} = -\frac{2}{-20} = +\frac{2}{20}$$

$$\therefore \quad \frac{1}{v} = -\frac{1}{20} \qquad \therefore \quad v = -20 \text{ cm (to left)} \quad Ans.$$

$$m = \frac{v}{u} = \frac{-20}{20} = -1 \quad \text{(Same size as object but inverted, and therefore image is real.)} \quad Ans.$$

Problem

Where would an object have to be located in front of a concave mirror in order to have a virtual image formed? Precisely where would it be located if the radius of the mirror were 20 cm and the image were 20 cm behind (i.e., to the right of) the mirror?

Solution

FORMULA METHOD

$$\frac{1}{u} - \frac{1}{v} = -\frac{2}{R}$$

$$\frac{1}{u} - \frac{1}{20} = -\frac{2}{-20} \qquad \therefore \quad \frac{1}{u} = \frac{3}{20} \qquad \therefore \quad u = 6\frac{2}{3} \text{ cm in front of mirror.} \quad Ans.$$

RAY-DIAGRAM METHOD

To form a virtual image with a concave mirror, the object must lie between the focal point and the mirror.

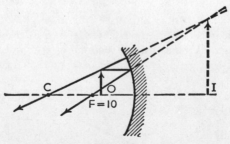

Fig. 110

For an image to be 20 cm to the right of the mirror, the ray approaching the head end of the image from the focal point must have been reflected through the focal point, whereupon it must have been parallel to the axis before reflection.

The ray approaching the head end of the image from the center of curvature must have experienced no change in direction.

Consequently, by tracing these two rays back to their origin, the head end of the object is located as in the diagram, at approximately $6\frac{2}{3}$ cm in front of the concave mirror. The diagram checks the calculations.

Problem

An object in the form of an arrow stands 1.2 cm high 24 cm in front of a convex mirror whose radius of curvature is 16 cm. What is the nature of the image? Where is it? How large is it?

Solution

Formula Method

$$\frac{1}{u} - \frac{1}{v} = -\frac{2}{R}$$

$$\frac{1}{24} - \frac{1}{v} = -\frac{2}{16}$$

$$\therefore \quad \frac{1}{v} = \frac{1}{24} + \frac{1}{8} = \frac{1}{24} + \frac{3}{24} = \frac{4}{24}$$

$$v = \frac{24}{4} = +6$$

Image is 6 cm to the right of the surface.   *Ans.*

$$m = \frac{v}{u} = \frac{6}{24} = \frac{1}{4}$$

Image is erect (virtual) and is $\frac{1}{4}$ as large as the object, or .3 cm.
    Ans.

Ray-Diagram Method

Image is approximately 6 cm to right of the surface and is approximately $\frac{1}{4}$ the size of the object. See Fig. 111.

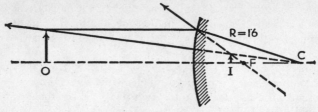

Fig. 111

Problem

This problem is an application of the single surface formula to refraction. A glass marble 2 cm in diameter has a small flaw .5 cm from the surface. Where does it appear to be, and how large does it appear to be with respect to its actual size?

Solution

This is a direct application of the paraxial-ray formula. Draw a diagram, noting that the flaw (presumed illuminated) is the object from which light passes from glass ($n_1 = 1.5$) to air ($n_2 = 1.0$).

$D = 2$ cm $\therefore$ $R = 1$ cm
$u = .5$ $v = ?$

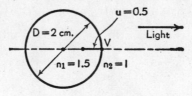

Fig. 112

$$\frac{n_1}{u} + \frac{n_2}{v} = \frac{n_2 - n_1}{R}$$

$$\frac{1.5}{.5} + \frac{1}{v} = \frac{1.0 - 1.5}{-1}$$

$$\frac{1}{v} = .5 - 3 = -2.5$$

$$v = -\frac{1}{2.5} = -.4 \quad \text{(Image is .4 cm to left of V.)}$$

$$m = -\frac{n_1 v}{n_2 u} = -\frac{1.5(-.4)}{1(.5)} = +1.2 \quad \text{(Image is erect, virtual, and 1.2 times as large as object.)} \quad \textit{Ans.}$$

Problem

A flat-bottom swimming pool is 8 ft deep. How deep does it appear to be when filled with water whose refractive index is $\frac{4}{3}$?

Solution

This is a straightforward application of the single surface formula for a surface whose radius of curvature is infinite.

The illuminated bottom of the pool (or a small portion of it) is the source of light which passes from water $(n_1 = \frac{4}{3})$ to air $(n_2 = 1.0)$. The object distance is 8 ft. The problem is to determine the image distance.

To be consistent with the conventions introduced earlier, the pool should be drawn on edge to allow light to pass from left to right, i.e., from medium n_1 to medium n_2. See Fig. 113.

Apply the single surface formula

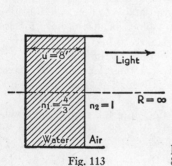

$$\frac{n_1}{u} + \frac{n_2}{v} = \frac{n_2 - n_1}{R}$$

$$\frac{\frac{4}{3}}{8} + \frac{1}{v} = \frac{1 - \frac{4}{3}}{\infty} = 0$$

$$\therefore \frac{1}{v} = -\frac{4}{24} = -\frac{1}{6}$$

$$\therefore v = -6$$

Fig. 113

(Image is on same side as object. The pool appears to be 6 ft deep instead of 8 ft.) *Ans.*

Problems: Single Spherical Surface

1. If the focal length of a concave mirror is 60 cm, what is its radius of curvature?

2. If an object is placed 50 cm in front of a concave mirror of 60 cm radius, where is the image? State distance of image from mirror and whether in front of or in back of mirror.

3. Given a spherical mirror whose radius of curvature is +20 cm. Sketch this mirror and indicate the position of the focal point. (a) How long is the focal length of this mirror? (b) Calculate the image distance for an object 30 cm in front of the mirror. (c) Is the image in front of the mirror or behind it? (d) Is the image real or virtual?

4. A concave mirror of radius 60 cm is so placed that a luminous object is 20 cm from the vertex. (a) Locate the image. (b) Is it virtual or real?

5. A point source of light is 3 ft below the surface of a pond of water, index of refraction = 1.33. Find the sine of the minimum angle of incidence for total internal reflection.

6. The front surface of a block of glass ($n = 1.5$) is spherical with a radius of curvature of 15 cm and is convex. If an object is placed 5 cm in front of the surface, what is the image distance? Is the image in front of or behind the surface?

7. A man 6 ft tall stands 10 ft in front of a bull's eye (convex) mirror which has a radius of curvature of 5 ft. (a) Is the image real or virtual? (b) How tall is the image? (c) Is the image on the same side of the mirror as the man? (d) Is the image inverted or erect?

8. A photographer takes a picture of a very clear pool of water from directly above it. He adjusts his camera so as to focus on the bottom. If the pool is actually 4 ft deep, what is the apparent depth? The index of refraction for water is $\frac{4}{3}$.

9. A small flaw appears when viewed from above to be located inside of and 5 cm beneath the surface of a block of glass whose top is flat. Assuming the refractive index of glass to be 1.6, calculate by the single surface formula the actual depth of the flaw.

10. A person is looking into the side of a rectangular fish tank filled with water whose index of refraction is 1.33. The back of the tank appears to be 3 ft back of the front surface. How far is it?

11. A cat looks at a goldfish which is at the center of a spherical bowl filled with water. The radius of the bowl is 20 cm. (a) Where is the image of the fish? Neglect the effect of the glass. The index of refraction for water is $\frac{4}{3}$. (b) What is the magnification of the image?

12. What is the radius of curvature of a convex mirror which forms an image one fourth the size of an object, the latter being 6 ft in front of the mirror?

13. Where must an object be placed in front of a concave mirror of radius R in order that the image be superimposed upon the object? Is the image real or virtual?

14. An ornamental silvered ball 6 in. in diameter forms an image of an object 2 ft in front of the ball. Locate the image by a diagram and calculate its position.

15. Sunlight shines on a glass sphere of radius 4 in. Using the single surface formula for each surface in succession, treating the image from the first surface as the object for the second surface, calculate

the position of the image of the sun ($n = 1.5$). *Hint!* Whenever an image from a preceding surface acts as the object for a second surface, but is located to the right of the second surface, it is called a virtual object and must be designated with a negative sign.

16. A person standing 12 in. in front of a concave mirror of 36 in. radius sees an image of his own eye. Calculate the location and the magnification of the image.

17. A plano-convex glass lens 2 in. thick lies flat side down on a newspaper. The curved surface has a radius of curvature of 12 in., and the glass has a refractive index of 1.50. Locate and determine the magnification of the image of the newsprint in contact with the glass surface, when it is viewed from above.

18. A polished glass cube 2 in. to a side lies on a sheet of newsprint. Where does the image of the newsprint appear to be for an observer directly above the cube ($n = 1.5$)?

19. If a concave mirror of radius −3 ft is used to project on a screen an image of an object with a magnification of 3, how far from the mirror must the object be placed?

20. Using the single surface formula, calculate the location and size of the image of a 6 ft person standing 3 ft in front of a plane mirror.

The Thin Lens in Air. The methods just outlined for locating the image formed by a spherical surface can be used for a succession of centered spherical surfaces. The image formed by the first one becomes the object for the second one, etc. On the other hand, it is more convenient, in the case of two surfaces rather close together with the same first and third media (a *thin lens*), to solve the problem algebraically for a single formula, rather than to solve the problem of each surface separately.

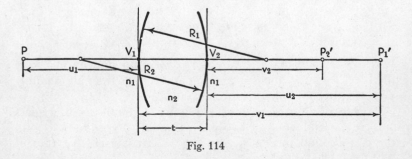

Fig. 114

In Fig. 114, u_1 is the object distance from the vertex V_1 of the first surface, whose radius of curvature is R_1 and which separates medium n_1 from medium n_2. Application of the paraxial-ray single-surface formula:

$$\frac{n_1}{u_1} + \frac{n_2}{v_1} = \frac{n_2 - n_1}{R_1}$$

yields a value for v_1. When the first image is calculated to lie to the right of the vertex v_2 of the second surface, whose radius of curvature is R_2, the object for the second surface obviously cannot be taken to the left of it as convention requires. By giving u_2 a negative sign, however, the corresponding formula can be used for the second surface:

$$\frac{n_2}{u_2} + \frac{n_1}{v_2} = \frac{n_1 - n_2}{R_2}$$

Thereupon v_2 can be calculated.

By combining these two equations algebraically, and noting that $u_2 = -v_1$ if the separation t is negligible, a single formula is obtained which, in the case of two surfaces close together in air, becomes

$$\frac{1}{u} + \frac{1}{v} = (n - 1)\left(\frac{1}{R_1} - \frac{1}{R_2}\right) = \frac{1}{f}$$

where n is the refractive index of the lens and f is the focal length of the lens. This is known as the *thin lens* (in air) *formula*.

Similarly the formulas for magnification at each surface, when combined, yield

$$m = m_1 \times m_2 = -\frac{v}{u}$$

It is to be noted that the value of v corresponding to $u = \infty$ is the focal length f. In other words, parallel rays from an infinitely distant source are "focused" at a point.

Graphical Considerations. Treating a double convex glass lens as a combination of two prisms mounted base-to-base with the corners smoothed off as in Fig. 115, it is seen that rays of light parallel to the axis of the lens (common base of the prisms) are so bent by refraction (toward the normal at the air-to-glass surface, and away from the normal at the glass-to-air surface) as to come to a common point F. This type of lens is called a *converging* lens.

On the other hand, a double concave glass lens causes light rays parallel to the axis to behave as in Fig. 116. This type of lens

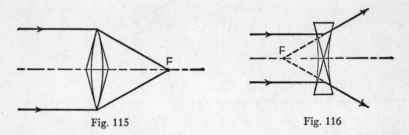

Fig. 115   Fig. 116

is called a *diverging* lens. It has a virtual focal point F and therefore a negative focal length in the lens formula.

At the center of a thin lens, the surfaces are approximately parallel. When a ray of light is obliquely incident upon a glass block having plane and parallel surfaces, it is refracted as in Fig. 117. The bending toward the normal at the air–glass surface is nullified by the bending away from the normal at the glass–air surface, so that the emerging ray is parallel to the incident ray but displaced by an amount which is proportional to the thickness of the block, i.e., the separation of the parallel surfaces. Obviously if the separation is negligible as in the case of a thin lens, the displacement is negligible. Thus any ray which passes through the center of a thin lens is undeviated in direction.

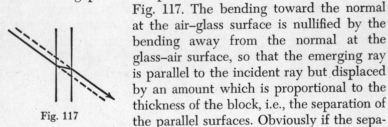

Fig. 117

Ray Diagrams for Lenses. The above analyses of rays of light with respect to prisms and planes can be utilized to locate images of objects as they are produced by lenses. The procedure is as follows.

CONVERGING LENS WITH OBJECT OUTSIDE FOCAL POINT (Fig. 118).

1. Draw a diagram as in the case of curved mirrors with the

object to the left and on the axis of the lens. Specify the focal points (note that there are two of them) from the known focal length of the lens.

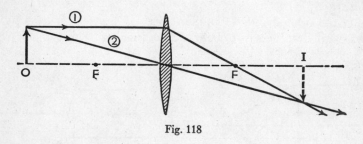

Fig. 118

2. From the head end of the object, select the ray parallel to the axis and direct it through the opposite focal point after refraction.

3. Pass a ray from the head end of the object directly through the center of the lens. This ray is undeviated by refraction.

4. The point of intersection of these two rays after refraction is a reproduction (image) of the point of their origin (object).

DIVERGING LENS. Follow the same procedure as in the case of the converging lens, except that the ray parallel to the axis, instead of passing through the opposite focal point after refraction, is diverged so as to appear to come from the near focal point as in Fig. 119. In this case the image is always virtual, lies to the left of the lens, and is reduced in size compared with the object. Moreover, it is always erect.

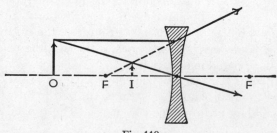

Fig. 119

CONVERGING LENS WITH OBJECT INSIDE FOCAL POINT. As in the case of the other converging system, the concave mirror, the nature and location of the image of an object depends upon the location of the object with respect to the focal point. If the object is closer to the lens than the focal point, the image is virtual, erect, on the same side of the lens as the object, and enlarged. See Fig. 120.

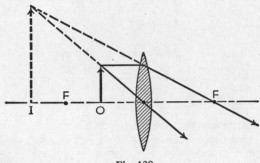

Fig. 120

Problem Procedures

Always draw a ray diagram, if for no other reason than to check the calculations from the lens formula, both for the location and the nature of the image. Numerical problems involving lenses are but applications of the formulas

$$\frac{1}{u} + \frac{1}{v} = \frac{1}{f} = (n - 1)\left(\frac{1}{R_1} - \frac{1}{R_2}\right)$$

and

$$m = -\frac{v}{u}$$

It is to be noted that the formulas for spherical mirrors and the formulas for thin lenses are simply adaptations of the single spherical-surface paraxial-ray formulas. They are not to be thought of as independent formulas involving separate conventions.

Problem

An object in the form of a vertical arrow 8 cm high is 24 cm in front of, and on the axis of, a double convex lens of 16 cm focal length. Locate the image. Is it real or virtual, and how large is it?

Solution

By the formula:
$$\frac{1}{u} + \frac{1}{v} = \frac{1}{f}$$
$$\frac{1}{24} + \frac{1}{v} = \frac{1}{16}$$
$$\frac{1}{v} = \frac{1}{16} - \frac{1}{24} = \frac{3}{48} - \frac{2}{48} = \frac{1}{48}$$
$$\therefore \quad v = +48$$

The image is 48 cm from the lens and on the opposite side from the object.   *Ans.*

$$m = -\frac{v}{u} = -\frac{48}{24} = -2$$

The minus sign means that the image is inverted and therefore is *real*. The 2 means that the image is twice as large as the object, i.e., it is 16 cm high.   *Ans.*

RAY DIAGRAM

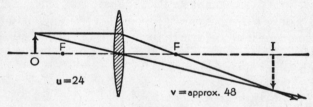

Fig. 121

Problem

If in the above problem the object is advanced to a position 8 cm in front of the lens, where and how large is the image?

Solution

Again by the formula:
$$\frac{1}{u} + \frac{1}{v} = \frac{1}{f}$$
$$\frac{1}{8} + \frac{1}{v} = \frac{1}{16}$$
$$\frac{1}{v} = \frac{1}{16} - \frac{1}{8}$$
$$\frac{1}{v} = \frac{1}{16} - \frac{2}{16} = -\frac{1}{16}$$
$$\therefore \quad v = -16 \text{ cm}$$

This means that the image is 16 cm in front of the lens.   *Ans.*

$$m = -\frac{v}{u} = +\frac{16}{8} = +2$$

This means that the image is erect, therefore virtual, and twice as large as the object, or 16 cm high.   *Ans.*

RAY DIAGRAM

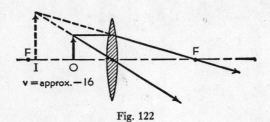

Fig. 122

Problem

Suppose, in the preceding two problems, the lens had been double concave with focal length 16 cm. Locate and determine the size of the image in each case, i.e., (a) the object at 24 cm, and (b) the object at 8 cm.

Solution

(a) If $u = 24$ cm: $\dfrac{1}{u} + \dfrac{1}{v} = \dfrac{1}{f}$ $\dfrac{1}{24} + \dfrac{1}{v} = \dfrac{1}{-16}$

$$\frac{1}{v} = -\frac{1}{16} - \frac{1}{24} = -\frac{3}{48} - \frac{2}{48} = -\frac{5}{48}$$

$$\therefore \quad v = -\tfrac{48}{5} = -9.6 \text{ cm}$$

$$m = -\frac{v}{u} = +\frac{9.6}{24} = +.4$$

These answers mean that the image is 9.6 cm in front of the lens, is erect and therefore virtual, and is .4 as large as the object, i.e., 3.2 cm high.   *Ans.*

Ray Diagram

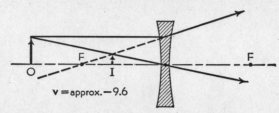

$v = $ approx. -9.6

Fig. 123

(b) If $u = 8$ cm:

$$\frac{1}{u} + \frac{1}{v} = \frac{1}{f} \qquad \frac{1}{8} + \frac{1}{v} = \frac{1}{-16}$$

$$\frac{1}{v} = -\frac{1}{16} - \frac{1}{8} = -\frac{1}{16} - \frac{2}{16} = -\frac{3}{16}$$

$$\therefore \quad v = -\frac{16}{3} = -5\frac{1}{3} \text{ cm}$$

$$m = -\frac{v}{u} = +\frac{\frac{16}{3}}{8} = \frac{2}{3}$$

These answers mean that the image is $5\frac{1}{3}$ cm to the left of the lens, is erect and therefore virtual, and is $\frac{2}{3}(8)$ cm high, or 5.33 cm.   *Ans.*

Ray Diagram

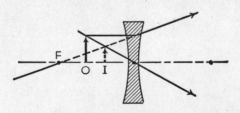

Fig. 124

Problem

If the double concave lens of 16 cm focal length in the preceding problem is symmetrical and is made of glass of refractive index 1.5, what are the radii of curvature?

Solution

Using the so-called lens-maker's formula

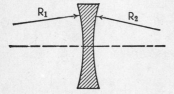

Fig. 125

$$\frac{1}{f} = (n-1)\left(\frac{1}{R_1} - \frac{1}{R_2}\right)$$

$$\frac{1}{16} = (1.5 - 1)\left(\frac{1}{R_1} - \frac{1}{R_2}\right)$$

$$\therefore \left(\frac{1}{R_1} - \frac{1}{R_2}\right) = \frac{\frac{1}{16}}{\frac{1}{2}} = \frac{1}{8}$$

In this case R_1 is negative, and R_2 is positive.

$$\therefore \quad \frac{1}{-R_1} - \frac{1}{R_2} = -\frac{2}{R_2}$$

$$\therefore \quad -\frac{2}{R} = -\frac{1}{8}$$

$$R = 16$$

This means that $R_1 = -16$ cm, and $R_2 = +16$ cm.   *Ans.*

Lens Power. It is to be noted that the shorter the focal length of a lens, the more refractive power the lens has, i.e., the more the light rays are bent. The reciprocal of the focal length is defined as the *power* (P) of the lens. The power is expressed in *diopters* when the focal length is expressed in meters. Converging lenses have *positive* power and diverging lenses have *negative* power.

Problem

If a spectacle lens has a specified power of $\frac{3}{4}$ diopters, what is its focal length?

Solution

A power of $\frac{3}{4}$ diopters means a focal length of $\frac{4}{3}$ meters or 133.3 cm.   *Ans.*

Problem

A so-called convexo-concavo lens has radii of curvature as follows:

$$R_1 = +70 \text{ cm} \quad \text{and} \quad R_2 = +50 \text{ cm}$$

What is the power of this lens in diopters? Assume $n = 1.5$.

Solution

Draw a diagram.

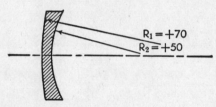

$R_1 = +70$
$R_2 = +50$

Fig. 126

This lens is obviously thicker at the rim than at the center, suggesting that it is a diverging (negative) lens.

Check by the formula

$$\frac{1}{f} = (n-1)\left(\frac{1}{R_1} - \frac{1}{R_2}\right)$$
$$= (1.5 - 1)\,(\tfrac{1}{70} - \tfrac{1}{50})$$
$$= .5\,\frac{(50-70)}{(3500)} = \frac{1}{2}\frac{(-20)}{(3500)} = -\frac{1}{350}$$
$$\therefore \quad f = -350 \text{ cm} \quad \text{or} -3.5 \text{ m}$$
$$P = \frac{1}{f} = -\frac{1}{3.5} = -.286 \text{ diopters} \qquad Ans.$$

Problems: Thin Lens

1. A double convex (converging) lens has a focal length of +30 cm. (a) Calculate the image distance for an object placed 20 cm in front of the lens. (b) Is the image in front of or beyond the lens? (c) What is the magnification? (d) Is the image real or virtual?
2. A converging lens has radii of curvature of +40 cm and −30 cm respectively and a refractive index of 1.5. What is its focal length in air?

3. A lens has a power of -4 diopters. Express its focal length in centimeters and indicate whether it is converging or diverging.

4. An object 1 cm high is placed on the principal axis 5 cm to the left of a thin lens with a focal length of $+10$ cm. (a) Where is the image formed? (b) How large is the image? (c) Is the image erect or inverted? (d) Is the image real or virtual?

5. If an object 10 cm in front of a diverging lens were 3 cm tall, where would the image be if it were 1.5 cm tall and erect? (State distance and position with respect to the lens.)

6. An object is placed 10 cm in front of a converging lens of 15 cm focal length. Where is the image? (Indicate whether in front of or behind the lens and how far.)

7. A plano-convex lens has a radius of curvature of 20 cm for the convex side. If the index of refraction is 1.5, what is the focal length of the lens?

8. A thin double convex lens ($R_1 = +50$ cm; $R_2 = -50$ cm) is made of material whose refractive index is 1.80. (a) What is the power of this lens in diopters? (b) Locate the image of an object placed 50 cm in front of this lens. (c) What is the magnification? (d) Is the image erect or inverted?

9. An object 6 cm high is placed 30 cm in front of a double convex lens of 25 cm focal length. (a) Sketch a ray diagram locating the image. (b) Find the position of the image by equation. What is the image distance? Is it behind or in front of the lens? (c) What is the size of the image? (d) Is the image real or virtual? Erect or inverted?

10. A diverging lens has radii of curvature of -30 cm and $+40$ cm respectively and a refractive index of 1.5. What is its focal length in air?

11. To what radius of curvature would the sides of a thin double convex lens have to be ground for the focal length to be 50 cm ($n = 1.5$)?

12. A rather satisfactory lens for vertical projection can be made by partially filling a watch glass with liquid. If water is used in a watch glass of radius R, how does the focal length depend upon the radius of curvature? Take the index of refraction of water to be 1.33.

13. An object 2.0 cm high is placed 30 cm in front of a converging lens of 15 cm focal length. (a) Compute the image position. (b) Compute the size of the image. (c) Is the image real or virtual? (d) What is the power of the lens in diopters?

14. A thin diverging lens ($R_1 = -50$ cm; $R_2 = +50$ cm) is made of glass whose refractive index is 1.60. Calculate (a) the position of

an image of an object 20 cm in front of the lens, (b) the magnification, (c) the diopter power of the lens.

15. A plano-convex glass lens ($n = 1.5$) is used to project on a screen an image 4 times the size of the object. If the object distance is 10 cm, what is the radius of curvature of the convex face of the lens?

16. An object 3 cm high is placed 20 cm in front of a lens whose power is +4 diopters. Where is the image and how large is it? Is the image real or virtual, erect or inverted?

17. Given a lens such that $R_1 = +20$ cm and $R_2 = +15$ cm made of glass of refractive index 1.50. An object is placed 30 cm in front of it. Locate the image by calculations.

18. A diverging lens has a power of −5 diopters. If an image is formed 10 cm from this lens, where must the object have been?

19. Carbon bisulfide has a refractive index of 1.63. Compare the diopter power of a lens of this material, made by filling a thin-walled double-convex mold whose sides have a radius of 20 cm, with one made by filling the mold with water of index 1.33.

20. A lens focuses sunlight on a screen at a distance of 30 cm from itself. By how much must the screen be moved to catch the image of an object as the object is moved from 5000 cm away up to only 50 cm from the lens?

Lenses in Combination. The mathematical and graphical methods of solving lens problems just described hold also for lens combinations, providing the lenses are centered on the same axis and provided they are reasonably close together. In such cases, the image produced by the first lens is used as the object for the second one, etc. If, as in the case of multiple surfaces, the image from the first lens lies beyond (to the right of) the second lens, it is treated as a *virtual* object for the second lens. This calls for a negative object distance in the formula and for special care in the ray-diagram method. Basically, however, the procedure follows the same pattern as that used for the single lens.

Problem

A double convex lens of focal length 20 cm is followed at a distance of 10 cm by a second double convex lens of focal length 20 cm. Locate and describe the nature and size of the image formed by the combination of an object 8 cm high placed 30 cm in front of, and on the axis of, the first lens.

Solution

This time consider the ray-diagram method first to get a general idea of the solution.

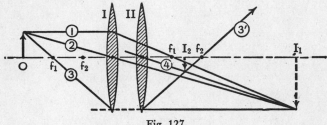

Fig. 127

Draw (1) the parallel ray through f_1, and draw (2) the ray through the center of the first lens to locate the image I_1 by lens I. Actually, of course, this image is prevented by lens II.

Now, of the rays of light tending to form image I_1, one ray can be found parallel to the axis. This one, designated as ray (3) in Fig. 127, must have come through the focal point f_1. As this ray strikes lens II, it must be diverted through the focal point f_2, becoming ray (3′), because it was parallel to the axis before it encountered lens II.

Moreover, one ray of those tending to form image I_1 must have come through the center of lens II. This is designated as ray (4) in Fig. 127. Because it passes through the center of lens II it is undeviated, and intersects ray (3′) so as to form image I_2, which is the so-called final or actual image formed by the system.

Graphically, the image is approximately 13 cm from the second lens, is inverted and therefore real, and is approximately 4 cm large.

FORMULA METHOD

For the first lens

$$\frac{1}{u_1} + \frac{1}{v_1} = \frac{1}{f_1}$$

$$\frac{1}{30} + \frac{1}{v_1} = \frac{1}{20}$$

$$\frac{1}{v_1} = \frac{1}{20} - \frac{1}{30} = \frac{30 - 20}{600} = \frac{10}{600} \qquad \therefore \quad v_1 = +60$$

$$m = -\frac{v_1}{u_1} = \frac{-60}{30} = -2$$

These answers mean that image I_1 is 60 cm to the right of lens I is inverted, is real, and is twice the size of the object, or 16 cm.
For the second lens:

$$\frac{1}{u_2} + \frac{1}{v_2} = \frac{1}{f_2}$$

$$\frac{1}{-50} + \frac{1}{v_2} = \frac{1}{20}$$

$$\frac{1}{v_2} = \frac{1}{20} + \frac{1}{50} = \frac{50 + 20}{1000} = \frac{70}{1000}$$

$$\therefore \quad v_2 = + \frac{100}{7} = +14.3$$

$$m = - \frac{v_2}{u_2} = - \frac{14.3}{-50} = +.29 \text{ approx.}$$

These answers mean that the final image is 14.3 cm beyond lens II, and is erect with respect to image I_1, which itself is inverted and real. Therefore the final image I_2 is inverted and real. It is approx. .29 the size of image I_1, which itself is twice the size of the object. Therefore the final image is .29(2) = .58 the size of the object, or .58(8) = 4.64 cm high.   *Ans.*

Problem

A certain double convex lens has a focal length of 20 cm. In combination with a second lens, the pair have a power of 4 diopters. What is the focal length of the second lens? Is the second lens a converging or a diverging lens?

Solution

The powers of lenses are additive.

$$P = P_1 + P_2$$

But
$$P_1 = \frac{1}{f} = \frac{1}{20} = 5 \text{ diopters}$$

$$\therefore \quad 4 = 5 + P_2$$

$$P_2 = -1 \text{ diopter}$$

$$f_2 = \frac{1}{P_2} = - \frac{1}{1} = -1 \text{ m}$$

$$= -100 \text{ cm.} \quad \textit{Ans.}$$

The negative sign indicates that the lens is a diverging lens.   *Ans.*

Optical Instruments. The fact that lenses and mirrors form optical images makes possible the construction of optical instruments, which can be used as visual aids. If an object cannot conveniently be observed because of its location or size, an image of appropriate size may be formed at a more convenient location by some sort of optical instrument. Or an image of an object may be formed on a photographic film where it may be preserved by suitable chemical action. Even the eye is an optical instrument whose function in part is to form on the retina an image of the object being observed. In general, optical instruments are lenses or mirrors or combinations of one or more of each. Sometimes prisms are incorporated because of their ability to change the direction of light rays.

Several optical instruments will be described and their ray diagrams will be treated as problems.

The Simple Microscope. It was shown earlier that a converging lens forms an upright, virtual, and magnified image of an object if the object lies closer to the lens than the focal point. Thus the simple lens used as a magnifier is to be thought of as an optical instrument, called the simple microscope. The reading glass used to magnify the printed page exemplifies the simple microscope.

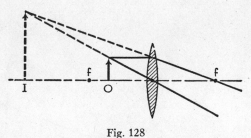

Fig. 128

The Astronomical Telescope. If an object lies beyond the focal point of a converging lens, the image is real and inverted and on the opposite side of the lens from the object. Thus to view the moon or some other luminous astronomical object which lies too far away for close examination, a real image of the object is first formed in the manner just mentioned, and then this real image is viewed through a second lens held so close to it that the image is within the focal length. The first lens is called the *objective lens,* and the second one is called the *ocular.* The combination gives an enlarged virtual image I_2 of the real image I_1 of the object O, as in Fig. 129.

The Compound Microscope. The compound microscope functions in a manner quite similar to that of the astronomical telescope, but in this case the objective has a very short focal

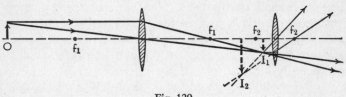

Fig. 129

length compared with the ocular. The objective of the astronomical telescope has a long focal length compared with the ocular.

The Terrestrial Telescope or Spyglass. Because the image formed by the astronomical telescope is inverted, this instrument is not ideally suited for viewing terrestrial objects. By the use of an intermediate lens, however, the image can be reinverted so that it is erect, as is seen in Fig. 130. The first image I_1 is made to lie outside the focal length of the second lens, which forms a

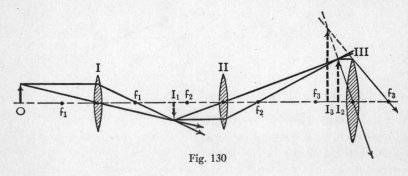

Fig. 130

real image I_2 of it beyond the lens. This second image I_2 is made to lie within the focal length of the third lens, which thereby forms a virtual image I_3 of it on the same side of the lens. I_3 is of course the final image (and the only actual image) formed by the system, and is erect. The primary disadvantage of this arrangement is the length of the resulting instrument, often called the spyglass.

The Galilean Telescope or Opera Glass. Galileo devised an ingenious way of producing an erect image by the use of only two lenses. The objective is a converging lens, but the ocular is a diverging lens. The operation of this instrument is seen from the ray diagram in Fig. 131. The image I_1 which would be formed by lens I alone is not allowed by lens II, which, being a negative lens, diverges the rays headed for I_1 and makes them appear to have come from image I_2. This image I_2 is a virtual image and is erect. In practice the separation of the lenses is usually less than

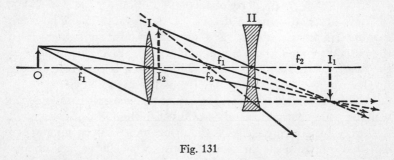

Fig. 131

indicated in the diagram, which was purposely exaggerated to make the ray tracing clear.

The Camera. The photographic camera is essentially a light-tight box with a converging lens system mounted in a hole in the front and a receiving screen for the image mounted in the rear. The objective lens forms a real image of the object being photographed. This image is made to fall on the screen, which is a film of light-sensitive material, whereupon it is preserved by later treatment with chemicals in a process called "development." See Fig. 132.

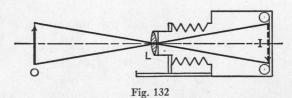

Fig. 132

The Eye and Its Defects. The eye operates very much like the camera box. A converging lens system forms a real image on the retina, which is a film of nerve tissue spreading out over the rear interior of the eyeball, constituting an extension of the optic nerve. The light pattern on the retina produced by the real image activates the nerve ends in the optic cord, which transmits a message to the brain. Thus vision is explained.

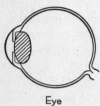

Eye

Fig. 133

The lens of the eye is so attached to muscles that it can be squeezed or stretched to change its focal length and thereby give the eye the property of accommodation.

There are two common defects of the human eye. These are *myopia* (near-sightedness) and *hyperopia* (far-sightedness). In the case of myopia, the introduction of a negative lens in front of the eye corrects the difficulty, and hyperopia is corrected by the similar use of a positive lens. Whereas the myopic eye can "see" only objects which are close by, due to excessive convergence of light, the negative lens tends to diverge the light or counteract the excessive convergence. On the other hand, the hyperopic eye can "see" only objects far away because of a relative lack of convergence, which is remedied by a supplementary converging lens.

Problem Situations Involving Spectacle Lenses. Obviously the function of a spectacle lens is to place an image of an object where the eye can see it. Normal reading distance is assumed to be 25 cm or 10 in.; if a myopic eye which, let us say, can see only at 15 cm, wishes to read a sign at 100 cm it requires a lens of such focal length as to form an image at 15 cm.

$$\frac{1}{u} + \frac{1}{v} = \frac{1}{f}$$

$$\frac{1}{100} + \frac{1}{-15} = \frac{1}{f}$$

where the minus sign indicates that the image is to be on the same side of the lens as the object.

$$\therefore \quad \frac{1}{f} = \frac{15 - 100}{1500} = \frac{-85}{1500}$$

$$f = -17.6 \text{ cm}$$

This means that a diverging lens of -17.6 cm focal length, or $1/.176$ or 5.7 diopters (negative), must be used.

Problem

A person who can see clearly nothing closer than 60 cm wishes to examine the details of an object 20 cm away. What kind of spectacle lenses and of what power must he use?

Solution

Obviously the person is far-sighted, since the normal reading distance for the average eye is approx. 25 cm. This suggests that spectacles be used of such focal length that an object distance of 20 cm will correspond to an image distance of -60 cm (minus sign indicates image on same side of lens as object).

$$\frac{1}{u} + \frac{1}{v} = \frac{1}{f}$$

$$\frac{1}{20} + \frac{1}{-60} = \frac{1}{f}$$

$$\frac{1}{f} = \frac{3}{60} - \frac{1}{60} = \frac{2}{60} \qquad \therefore \quad f = +30 \text{ cm (converging lens)} \qquad Ans.$$

The power of the lens is $1/.3 = 3.33$ diopters. *Ans.*

Problems: Lenses in Combination, Optical Instruments

1. Two converging lenses are located 20 cm apart. The first lens has a focal length of 40 cm, the second lens a focal length of 25 cm. Where will parallel light striking the first lens be brought to focus?

2. A simple magnifier is used to view an object 2 cm from the lens. What focal length must the lens have to form an image 25 cm from the lens?

3. A projection lantern is essentially a box containing a lens which projects on a distant screen an image of the lantern slide. If the projection lens to screen distance is 24 ft, and the slide to projection lens is 1 ft, how long will a line 1 in. long on the slide appear on the screen?

4. A thin lens of $+5$ cm focal length is placed in contact with another thin lens of -5 cm. (a) Find the focal length of the combination. (b) How would the focal length of such a combination compare with the focal length of a piece of glass with plane parellel sides? (c) How can you measure the focal length of a diverging lens? Describe fully.

5. A diverging and a converging lens are placed in contact. The focal

length of the combination is found to be +30 cm. If the focal length of the diverging lens is −20 cm, what must be the focal length of the converging lens?

6. Two converging lenses, A and B, are identical except that the index of refraction of A is 1.5, and that of B is 1.6. Which lens has the greater focal length?

7. A simple magnifier held close to the eye is used to view an object. What focal length must it have to form an image at an infinite distance from the lens if the magnifying power is 5 times? *Note!* It can be shown that when a single lens is used as a magnifier, the magnifying power is $m = 25/f$ if the image is at infinity, or $m = (25/f) + 1$ if the image is at 25 cm.

8. An object 2 cm high is placed on the axis and 20 cm in front of a converging lens whose focal length is 15 cm. If a second lens exactly like the first one is placed 10 cm beyond the first one, determine the location and size of the image formed by the combination.

9. A compound lens is made up of a converging lens of 20 cm focal length and a diverging lens of 30 cm focal length. (a) How far from an object must it be placed in order to produce a virtual image 25 cm from the lens? (b) What is the power of this lens?

10. A telescope objective forms an image of the moon 200 cm from itself in a darkened room. How close to this image must a magnifying glass be held to form a virtual image of this image 25 cm from the glass and magnified 6 times?

11. A certain far-sighted person cannot see distinctly an object closer than 70 cm. In order to see distinctly a printed page held at 25 cm, a spectacle lens of what power (sign as well as magnitude) must he use?

12. What must be the focal length of a slide projector lens to just "fill" a 5 × 5 ft screen at a distance of 10 ft using a 2 × 2 in. slide?

13. A simple magnifier has a focal length of 4 cm. (a) What is its magnifying power if the image is projected to infinity? (b) If projected to 25 cm? See note in Problem 7.

14. A photographic enlarger has a focal length of 4 in. At what distance from a 4 × 5 in. negative must it be placed to form an 8 × 10 in. enlargement?

15. A certain near-sighted person cannot see things distinctly farther than 6 in. away. If he uses a negative spectacle lens whose focal length is −7 in., at what minimum distance can an object be placed so that he can see it distinctly?

16. Two converging lenses are so mounted as to form a telescope. The objective has a focal length of 30 cm, and the eyepiece has a focal length of 1 cm. When used to view a distant object, the separation of the lenses is so adjusted that the image is projected to infinity, i.e., the separation is made equal to the algebraic sum of the focal lengths. (a) What is approximately the combined magnification when the instrument is focused on an infinitely distant object? (b) When it is focused on an object at 100 cm and the final image is formed at 25 cm from the eyepiece, what is the magnification?

17. Locate the position, size, and nature of the image formed when an object is placed 60 cm in front of a converging lens of focal length 40 cm, followed at 15 cm by a second converging lens whose focal length is 30 cm.

18. An opera glass has objective lenses of 4 in. focal length combined with ocular lenses of 1 in. focal length (negative). In use, they give a virtual image, 10 in. in front of the ocular lenses, of an object 6 ft high and 25 ft in front of the objectives. Calculate (a) the size of the virtual image, (b) the separation of the lenses.

19. A person's prescription for glasses reads "D = − 2 diopters." (a) What is the far point of this person's eye? (b) What would his near point be if the prescription had read "D' = +2 diopters"?

18

Physical Optics

A most important optical instrument is one which makes possible the analysis of the light given off by various sources. Although it might well have been discussed in the previous chapter as an instrument which forms optical images, it really introduces a whole series of phenomena known as spectroscopy.

The Spectroscope. The spectroscope is a device used for analyzing the component wave lengths present in a source of light. One form uses a prism which refracts light in such a manner that red light (long wave lengths) is bent through a smaller angle than blue light (short wave lengths). An illuminated slit serves as the object, which is imaged as a series of overlapping slits, one for each wave length present in the light source that is used. Thus so-called white light is spread out into a band of colors called the *spectrum* of white light.

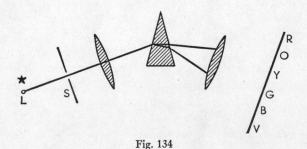

Fig. 134

By the use of the spectroscope, certain laws of spectroscopy have been formulated, but since these laws do not lend themselves to numerical problem situations, they will not be discussed here. Spectroscopy does, however, deal with the nature of light and therefore serves as a connection between geometric optics and physical optics. Although the latter is a very important aspect of optics, it will not be treated in such detail as the former because a quantitative treatment is beyond the mathematical level of a first course.

Phenomena of Physical Optics. The phenomena which characterize physical optics are interference, diffraction, and polarization. All three are allied with the wave nature of light. Just as water waves can be superimposed to yield reinforcement and nullification, so can light be made to interfere constructively and destructively. Hence the name *interference* is given to the phenomenon. Also, as water waves can bend around corners, so does light bend or change its direction as it passes close to the edge of fine slits or through fine gratings made by ruling lines close together on transparent or reflecting surfaces. This phenomenon is called *diffraction*. Certain crystalline substances like tourmaline cause light to vibrate in specific planes such that once it has passed through a sample of the substance, it can pass through a second sample only if the second is oriented the same way as the first. The name given to this phenomenon is *polarization*. These phenomena will now be discussed in turn, but very briefly because they do not lend themselves to many numerical problem situations within the scope of first-year Physics.

Interference. When light is incident upon very thin films of transparent materials, it may be reflected from both the front and the rear surfaces. If the thickness of the film is of the order of a few wave lengths of light, the two reflected beams may be in phase or out of phase with each other. Which they are depends upon whether the path difference is an even or an odd number of half wave lengths, and how the refractive index of the film compares with the refractive index of the medium in which it is found. Devices called interferometers are used to measure very small displacements. They utilize a pattern of interference fringes. As a displacement of one component of the optical system is produced, the number of fringes crossing a fiducial mark is counted, each one corresponding to a known fraction of a wave length of the light used. Thus a very precise measurement of a very slight displacement is possible.

Diffraction. Utilizing the wave nature of light, the direction of a beam of light is changed (the light is bent, so to speak) at the edges of obstructions. If light of a given wave length is passed through a grating, consisting of a great many lines ruled on a glass plate by a fine diamond point, it is bent through an angle θ according to the following relation:

$$n\lambda = d \sin \theta$$

where n is the "order" of the interference, λ is the wave length, and d is the separation of the rulings on the plate. Using such a "diffraction grating" the wave length of light can be measured to a high degree of precision.

Polarization. Light passed through a polarizing material is restricted to vibrate in a given plane. The effect can be observed by the use of a second polarizing material because the second sample must be oriented exactly like the first one to allow the light "polarized" by the first one to be transmitted through the second one. The first one is called the *polarizer,* and the second one is called the *analyzer.* The phenomenon of polarization indicates that if light is wavelike, the wave must be transverse rather than longitudinal.

Problems: Interference, Diffraction, and Polarization

1. Radiation of wave length 6×10^{-5} cm is diffracted by a grating having 5000 lines per cm. What will be the sine of the angle of deviation from the normal? (First-order diffraction.)
2. A certain diffraction grating has 14,600 lines to the inch (grating spacing of .000174 cm). Using green light of wave length .0000546 cm, what is the value of the first-order diffraction angle? Express the answer in terms of the sine of the angle.
3. A source of monochromatic light is placed 140 cm in front of a grating which has a grating space of .000174 cm (corresponds to 14,600 lines per inch). If the first-order image is 50.4 cm from the central image on a screen perpendicular to the undeviated beam, what is the wave length of the light?
4. A 500 lines per cm grating is placed 100 cm from a slit that is illuminated by filtered mercury light of wave length .00005461 cm. What is the distance from the central image to the second-order image on a screen perpendicular to the undeviated beam?

5. How many lines per centimeter must a grating have to produce first-order diffraction at an angle of 20° from the undeviated beam using light of wave length .0000589 cm?

Appendix

Symbols, Concepts, and Units

SYMBOL	CONCEPT	BRITISH ENG. UNIT (ft/slug/sec)	CGS UNIT (cm/g/sec)	MKS UNIT (m/kg/sec)
A	amplitude	ft	cm	m
a	acceleration (sometimes A)	ft/sec^2	cm/sec^2	m/sec^2
	area	ft^2	cm^2	m^2
B	bulk modulus	lb/ft^2	dyne/cm^2	newton/meter^2
	buoyant force	lb	dyne	newton
	magnetic induction		gauss or maxwell/cm^2	newton/amp m. or weber/m^2
b	internal battery resistance	ohm	ohm	ohm
C	compressibility	ft^2/lb	cm^2/dyne	meter^2/newton
	compression	lb	dyne	newton
	electrical capacitance		stat-farad	farad
°C	centigrade temperature			
c	heat capacity	Btu/lb/°F	cal/g/°C	cal/g/°C
d	density	slug/ft^3	g/cm^3	kg/m^3
	distance	ft	cm	m
E	electric field intensity		dyne/stat-coulomb	newton/coulomb
	electromotive force		stat-volt	volt
	illuminance	ft-candle	meter-candle	meter-candle
°F	Fahrenheit temperature			
f	focal length (mirror or lens)	ft	cm	m

SYMBOL	CONCEPT	BRITISH ENG. UNIT (ft/slug/sec)	CGS UNIT (cm/g/sec)	MKS UNIT (m/kg/sec)
	(sometimes F) force	lb	dyne	newton
g	acceleration of gravity	ft/sec^2	cm/sec^2	m/sec^2
H	heat	Btu	cal	cal
	horizontal force	lb	dyne	newton
	magnetic field intensity		dyne/pole = oersted	ampere/meter
h	height	ft	cm	meter
I	intensity or candle power	candle	candle	candle
	intensity of current	amp	ab-amp	amp
	moment of inertia	slug ft^2	g cm^2	kg m^2
K	dielectric constant		stat-coulomb^2/ dyne cm^2	coul^2/newton m^2
$°K$	absolute (Kelvin) temperature			
KE	kinetic energy	ft-lb	dyne-cm = erg	joule
k	constant of proportion- ality			
	electric resistivity	ohm ft	ohm cm	ohm m
L	coefficient of self-induct- ance		ab-henry	henry
	torque	lb ft	dyne cm	newton meter
l	(sometimes L) length	ft	cm	meter
	lever arm	ft	cm	meter
M	coefficient of mutual in- ductance		ab-henry	henry
	magnetic moment		pole cm	ampere meter^2
μ	magnetic permeability		pole^2/dyne cm^2	weber/amp meter

SYMBOL	CONCEPT	BRITISH ENG. UNIT (ft/slug/sec)	CGS UNIT (cm/g/sec)	MKS UNIT (m/kg/sec)
m	magnetic pole (sometimes		pole	
	M) mass	slug	gram	kg
mv	momentum	slug ft/sec	g cm/sec	kg m/sec
N	normal force	lb	dyne	newton
n	coefficient of friction	dimensionless		
	coefficient of rigidity	lb/ft^2	$dyne/cm^2$	$newton/m^2$
	frequency of vibration	number/sec	number/sec	number/sec
	index of refraction	dimensionless		
P	power	ft-lb/sec or hp	dyne-cm/sec	joule/sec or watt
	power of lens		diopter	diopter
PE	potential energy	ft-lb	dyne-cm = erg	joule
p	pressure	lb/ft^2	$dyne/cm^2$	$newton/m^2$
q	(sometimes Q) electric charge		stat-coulomb	coulomb
R	electrical resistance		ohm	ohm
	resultant force	lb	dyne	newton
r	(sometimes R) radius	ft	cm	meter
s	displacement	ft	cm	meter
	radiation constant	joule/sec/deg absolute to 4th power		
	(sometimes S) specific gravity	dimensionless		
	specific heat	dimensionless		
T	absolute temperature		°K	
	period of vibration	sec	sec	sec
	tension	lb	dyne	newton
t	temperature	°C, °F, °K		
	time	sec	sec	sec

SYMBOL	CONCEPT	BRITISH ENG. UNIT (ft/slug/sec)	CGS UNIT (cm/g/sec)	MKS UNIT (m/kg/sec)
u	object distance (mirror or lens)	ft	cm	meter
V	electric potential		stat-volt	volt
	vertical force	lb	dyne	newton
	volume	ft^3	cm^3	meter^3
v	image distance (mirror or lens) (sometimes V)	ft	cm	meter
	velocity	ft/sec	cm/sec	m/sec
W	weight	lb	dyne	newton
	work	ft-lb	dyne-cm $=$ erg	joule
X	reactance		ohm	ohm
Y	Young's modulus	lb/ft^2	dyne/cm^2	newton/m^2
Z	electro-chemical equivalent		g/coulomb	kg/coulomb
	impedance		ohm	ohm
α	angular acceleration	rad/sec^2	rad/sec^2	rad/sec^2
	linear co-efficient of thermal expansion	/°F	/°C	/°C
β	volume co-efficient of thermal expansion	/°F	/°C	/°C
θ	angle	radian	radian	radian
λ	wave length	ft	cm	meter
ϕ	angular displacement	radian	radian	radian
	luminous flux	lumen	lumen	lumen
	magnetic flux		line $=$ maxwell	weber
ρ	radius of gyration	ft	cm	meter
ω	angular velocity	radian/sec	radian/sec	radian/sec

Answers to Problems

1. (b) 41 lb
   (c) −8 lb
   (d) 42 lb
2. 150 lb
3. 10 lb
4. (a) 20 lb
   (b) 4 lb
   (c) .2

5. 1.5
6. 5 newtons, zero
7. 17.3 lb, 10 lb
8. Zero
9. 36 lb, 56° from A
   (approx)
10. Zero

1. (c) 577 lb
   (d) 1154 lb
2. (a) 40 lb
   (b) 17.3
   (c) .43
3. (c) 17.3 lb
   (d) 12.5 lb
   (e) Zero
4. (b) Up
   (d) 1.15 lb

5. (b) Down
   (d) 25 lb
   (e) .192
6. 93.1 lb, 132 lb
7. 69.3 lb
8. 250 lb
9. $C = 120$ lb, $T = 134$ lb
10. 4 lb

1. (a) 4330 lb ft
   (b) 17.3 ft
   (c) 34,600 lb ft
   (d) 38,900 lb ft
   (e) 1950 lb
2. (c) 135 lb
   (d) 450 lb ft
   (e) Less
3. (b) 86.6 lb
   (c) 866 lb ft
   (d) Greater
   (e) .866

4. (a) 1 lb
   (b) .376 lb
   (c) .376 lb
   (d) 1.07 lb
5. (b) 4.8 lb
6. 100 lb, 60 lb
7. 143.3 lb
8. Top: $V = 0$, $H = -45$
   lb; bottom: $V = 60$ lb,
   $H = +45$ lb
9. $V = 50$ lb up; $H =$ zero
10. $T = 86.6$ lb, $F = 132$ lb

261

Page 41

1. (a) 30 mi/hr
   (b) No
   (c) 4.4 ft/sec^2
   (d) 660 ft
2. 4 sec
3. 400 sec
4. (a) 19.4 ft/sec^2
   (b) 9.1 sec
5. 144 ft
6. 8.33 ft/sec^2 toward center

7. 176 meters
8. (a) 2 sec
   (b) 64 ft
   (c) 48 ft up
9. (a) 4.4 ft/sec^2
   (b) 660 ft
10. .297 meter/sec^2 toward center
11. 144 ft
12. 20 cm/sec^2

Page 51

1. 12 lb.
2. 2125 lb
3. 50 cm/sec^2 west
4. 10.7 ft/sec^2
5. (a) 5 slug
   (b) Down
   (c) 4 ft/sec^2 down
   (d) 160 lb

6. 19.1 meters/sec
7. .51 meter/sec^2 down
8. 87.9 newtons
9. 4.57 ft/sec^2, 4.57 ft/sec^2, 6.85 lb
10. 1173 lb

Page 57

1. (a) 20 ft/sec^2
   (b) 100 lb
   (c) 100 lb
   (d) .625
2. (a) .0001 slug
   (b) 8 ft/sec^2
   (c) .0008 lb
   (d) .25
3. (a) 44 ft/sec
   (b) 970 ft/sec^2
   (c) .69 slug ft/sec
4. (a) 6 ft/sec
   (b) 37.5 slug ft/sec

5. 8800 slug ft/sec
6. (a) 720 ft/sec
   (b) 960 ft/sec
   (c) 45 sec
   (d) 43,200 ft
7. 484 lb
8. $T = 59$ lb, $R = 5.94$ ft
9. 6 ft, 20 ft/sec at 53° with horizontal
10. 14 ft/sec
11. 4000 cm/sec

Page 63

1. (a) 26.2 meters/sec
   (b) 2.67 sec
   (c) 51.4 joules

   (d) 51.4 joules
   (e) 51.4 joules
2. 44 ft/sec

3. (a) 16,000 ft-lb
   (b) 9000 ft-lb
   (c) 7000 ft-lb
   (d) 20 lb
4. (a) 1.56 slug ft/sec
   (b) 117 ft-lb
   (c) 117 lb
5. (a) 123.7 ft/sec
   (b) 2.30 sec
   (c) 50 ft-lb

   (d) 59.8 ft-lb
6. 693 ft-lb
7. 2640 joules, .59 hp
8. (a) 3312 ft-lb
   (b) Zero
   (c) 11,250 ft-lb
   (d) 11,250 ft-lb
9. 2 ft
10. 387,200 ft-lb, 8.8 ft/sec², 70.4 hp

Page 67

1. 100
2. 16
3. 700 lb
4. (a) 5.0
   (b) 6.0
   (c) 83.3 percent

5. 60 lb, 120 lb
6. 30 lb at end
7. 300 lb
8. So as to get 5/1 ratio

Page 72

1. (a) 126 ft/sec
   (b) 351.7 rad
2. 750 rad
3. (a) 5 rad/sec
   (b) 47.7 rpm
   (c) 10 ft/sec

4. −10.47 rad/sec², 214 rev
5. (a) 2.09 rad/sec²
   (b) 20.9 rad/sec
6. (a) 25.1 rad/sec
   (b) 50.2 ft/sec
   (c) −.84 rad/sec²

Page 79

1. (a) 6 ft/sec²
   (b) 9.75 lb
   (c) 19.50 lb ft
   (d) 6.5 slug ft²
2. (a) 1.96 ft
   (b) 3.90 ft
   (c) 24 lb ft
   (d) 5 sec
3. (a) 15 cm/sec
   (b) 1.25 rad/sec
   (c) 225,000 erg
   (d) 78,100 erg
   (e) 30,000 g cm/sec
   (f) 125,000 g cm² rad/ sec

4. (a) 1 slug ft²
   (b) 12 slug ft² rad/sec
5. (a) 6.25 slug
   (b) 200 lb
   (c) .08 rad/sec²
   (d) 125 sec
6. .0523 newton
7. (a) .0587 slug ft²
   (b) .352 ft
   (c) −8.5 rad/sec²
   (d) 1.48 sec
   (e) 1.47 rev
8. 1.566 ft
9. 1.265 ft, .67 rad/sec²
10. 1.39 ft-lb

Page 84

1. (a) .1 sec
   (b) 5 inch
   (c) Down
2. (a) 6 cm
   (b) 113 cm/sec
   (c) Zero
   (d) 2150 cm/sec^2
   (e) 360 cm/sec^2
   (f) Zero
   (g) 6,450,000 dyne
3. (a) Zero
   (b) 62.8 cm/sec

   (c) 200 cm/sec^2
   (d) Zero
4. 4.2 sec
5. 986 cm/sec^2
6. 1.23 ft-lb
7. (a) 7.7 sec
   (b) 600 erg
   (c) 600 erg
8. 2.72 meters/sec;
   50 meters/sec^2; 200
   newtons

Page 92

1. (a) 41,600 lb/in.^{2}
   (b) .00083
2. (a) 50,000 lb/in.^{2}
   (b) 1500 lb

3. .062 cm
4. .085 lb/in.^{2}
5. 3.33 × 10^5 dyne/cm^2

Page 97

1. 170 cm
2. (a) 3333 lb
   (b) 33.3 lb/
       in.^{2}
   (c) 10 lb/
       in.^{2}

3. (a) 40 lb/
       in.^{2}
   (b) 400 lb
   (c) 5
4. 20,000 lb
5. 480 lb/ft^2

6. 9200 lb
7. 584,000 lb
8. 89.4 lb/in.^{2}
9. 1750 lb
10. 18,700 lb

Page 101

1. 3000 ton
2. 624 lb
3. 624 lb
4. 1000 cm^3

5. (a) .4 lb
   (b) 1.5 lb
6. .367
7. 187 lb

8. .7
9. 19.25 ft^3
10. .431 ft^3

Page 103

1. 66.7 ft^3
2. 350 ft^3

3. 44.1 lb/in.^{2}
4. 13,650 in.^{3}

5. 34 ft

Page 106

1. (a) 1.2 ft^2
   (b) 6 ft^3/sec

2. (a) 2496 lb/ft^2
   (b) 75 ft/sec

(c) 3 in.^{2} pipe

(d) 52 lb

3. 52.6 lb/in.^{2}, 4 times

Page 110

1. 22.2/sec
2. 4
3. 1056/sec
4. 14.4 × 10^9 dyne/cm^2
5. 267 cycle/sec

Page 115

1. 154 or 146
2. (a) 4 L
   (b) $\frac{4}{3}$ L
3. (a) 2 L
   (b) L
4. (a) 160 cm
   (b) 215/sec
   (c) 32 cm
5. 360 cm
6. 1.5 ft
7. 2200 ft
8. 1190 cycle/sec
9. (a) 442 cps
   (b) 357 cps
10. 13 cps

Page 124

1. 39° C
2. 8.33° C
3. .5 ft^3
4. .025 ft
5. 1.33 × 10^8 dyne/cm^2
6. 87.8° F
7. 5.54 ft
8. 1.32 cm^3
9. 30 lb/in.^{2}
10. 4.7 lb/in.^{2}

Page 130

1. (a) 18,000 cal
   (b) 606 cal
   (c) 634° C
2. 56.7 g
3. (a) 9000 cal
   (b) 303 cal
   (c) 580° C
4. 8.9° C
5. 95° C
6. 27.1° C
7. 3750 g
8. 15.75° C
9. 135,000 cal
10. 32.8 kg

Page 138

1. 2.21 × 10^6 cal
2. 50,000 cal/sec
3. 66 percent
4. 21 percent
5. 84 cal/min
6. 2.65 × 10^6 cal
7. 30,200 Btu
8. 1.66 watt
9. 266° C
10. 9200 ft-lb/sec

Page 146

1. (a) 100 dyne
   (b) 3 dyne/st-c toward 100
   (c) 13.3 cm
   (d) Zero
   (e) Toward 100

2. .6 st-c
3. 48 dyne
4. (a) 4 dyne/st-c
    (b) 8 dyne
    (c) Toward 400
    (d) 20 cm

5. 99.0 st-c
6. Zero
7. 120 dyne toward −600
8. 405,000 st-c
9. .108 newton, attraction

Page 156

1. 5 st-volt
2. 10 st-farad
3. (a) 120 st-volt
    (b) 40 st-volt
    (c) 160 erg
    (d) 160 erg
4. (a) 8 st-farad
    (b) 8 st-volt
    (c) 256 erg
5. (a) 80 st-volt
    (b) 80 st-volt
    (c) 160 erg
    (d) 160 erg
    (e) 1200 dyne
6. (a) 40 st-volt
    (b) 15 st-farad
    (c) 33.3 st-volt
    (d) 8340 erg
7. 8 st-farad
8. (a) 8 st-farad, 40 st-farad

    (b) 48 st-farad
    (c) 16 st-c
    (d) 96 st-c
9. (a) 2 st-farad
    (b) 4 st-c
    (c) 1 st-volt
10. (a) 4 st-farad
    (b) 4.8 st-c
    (c) 2.88 erg
    (d) 4.8 st-c
    (e) .8 st-volt
11. (a) 20 st-farad
    (b) 40,000 st-volt
    (c) 4 joule
    (d) 4 dyne/st-c
12. 925 st-volt
13. Zero
14. 3.08 farad
15. $V_A - V_B = 405$ volts
16. 2×10^{-5} coulombs

Page 171

1. 5 ohm
2. .25 amp
3. 7.5 volt
4. 30 volt
5. .3 amp
6. 250 coul
7. 75 volt
8. 500 watt
9. 20,000 cm
10. .269 g
11. 1500 joule
12. 48 watt

13. 1.7 ohm
14. (a) .1 ohm
    (b) 2.4 ohm
    (c) .9 watt
    (d) 21.6 watt
15. (a) .5 amp
    (b) 2.5 volt
    (c) 5 ohm
    (d) 20 ohm
16. 2,160,000 joule

17. (a) 3 amp
    (b) 3 amp
    (c) 1.2 volt
    (d) 1.2 volt
    (e) 1.5 joule
    (f) 3.6 watt
    (g) .9 watt
18. 2400 watt
19. (a) 300 amp
    (b) 3 volt
    (c) 3 volt
    (d) 60 joule

(e) 900 watt
(f) 6 volt
(g) Zero
20. $.50
21. 1.6 amp
22. (a) .13 amp
    (b) .91 amp
    (c) 885
      ohm
23. $.975

24. (a) 115 volt
    (b) 8.83
      amp
    (c) .87 amp
    (d) 221
      ohm
25. 1.58 volt
26. (a) 2.67
      volt
    (b) .667
      amp
27. 303 sec

28. 844 cal
29. 14.4° C
30. 1.5 volt,
    $R_a = .75$
    ohm,
    $R_b = 1.5$
    ohm

Page 179

1. 1.5 pole
2. 10 pole
3. 100 pole
4. 2 oersted
5. .156 oersted
6. 141.4 pole
7. 2.8 oersted, 8000 pole cm

8. (a) 8 oersted
    (b) 4000 gauss
9. 72°
10. .643 oersted, 70° below horizontal

Page 190

1. 12,500 dyne
2. 1.25×10^{-2} newton
3. 3.14×10^{-5} weber/m²
4. (a) 503 gauss
    (b) 603,600 gauss
    (c) 1,207,000 line (maxwell)

5. 2.4 dyne
6. 63°
7. .598 amp
8. 159 turns
9. 10 cm
10. .24 newton

Page 202

1. 3.77×10^{-5} volt
2. 5 henry
3. .0318 volt
4. $.5 \times 10^8$ ab-volt
5. 1 volt
6. 2860×10^{-8} volt
7. 10^7 ab-volt
8. 10^6 ab-volt
9. 200 cm/sec

10. .04 amp
11. 80 gauss
12. 3.08×10^{-5} amp
13. .094 volt
14. .4 volt
15. .00075 amp
16. 27.5 henry, 110 volt
17. .005 sec
18. 9.99×10^{-2} volt

Page 208

1. (a) 6 amp
   (b) 8.5 amp
   (c) Zero
2. $\frac{1}{3}$ amp
3. 50 ohm
4. (a) 10 amp
   (b) 14.1 amp
5. .725
6. 1.045 amp
7. 2 amp

8. (a) 11 ohm
   (b) 3.34 ohm
   (c) 2.78×10^{-2} henry
9. (a) 2 amp
   (b) 110 volt
   (c) 4400 watt
10. 92×10^3 cps
11. .87
12. 55 ohm

Page 214

1. 4 ft-candle
2. 3.6 ft-candle
3. 46.1 candle
4. 10 candle
5. (a) 600 lumen
   (b) 47.7 candle
   (c) 1 ft-candle
6. (a) 90 candle
   (b) $\frac{1}{40}$ ft-candle
7. 57,600 lumen
8. 956 candle

9. 8.26 ft-candle
10. 3 ft
11. 62.8 ft-candle
12. 69 candle
13. 4.87 ft-candle
14. 1.25 ft-candle
15. 13.9 ft-candle
16. 232 candle
17. 1800 candle
18. 64 lumen

Page 219

1. $\frac{9}{16}$
2. .798
3. 1.67
4. 1.49
5. 5.41×10^{14}/sec
6. .459
7. (a) 4 ft behind
   (b) 6 ft

   (c) 3 ft
   (d) No
8. 45°
9. 1.25×10^{10} cm/sec
10. 1.525
11. 52.6 cm
12. 63°

Page 229

1. 120 cm
2. 75 cm in front
3. (a) 10 cm
   (b) 7.5 cm

   (c) Behind
   (d) Virtual
4. 60 cm beyond, virtual
5. .75

6. 9 cm in front
7. (a) Virtual
   (b) 1.2 ft
   (c) No
   (d) Erect
8. 3 ft
9. 8 cm
10. 3.99 ft
11. (a) At center
    (b) None
12. 4 ft

13. At center of curvature, real
14. 1.41 in. below surface
15. 2 in. beyond sphere
16. 36 in. beyond, 3 times
17. 1.41 in. beneath top surface, 1.055
18. $1\frac{1}{3}$ in. below top surface
19. 1 ft in front
20. 3 ft behind, 6 ft high

Page 240

1. (a) -60 cm
   (b) In front
   (c) 3
   (d) Virtual
2. 34.3 cm
3. 25 cm, diverging
4. (a) 10 cm to left
   (b) 2 cm
   (c) Erect
   (d) Virtual
5. 5 cm in front
6. 30 cm in front
7. $+40$ cm
8. (a) 3.2
   (b) 83.3 cm beyond lens
   (c) 1.67
   (d) Inverted
9. (b) 150 cm behind
   (c) 30 cm

   (d) Real, inverted
10. -34.3 cm
11. 50 cm
12. $f = 3R$
13. (a) 30 cm beyond lens
    (b) 2 cm
    (c) Real
    (d) 6.67
14. (a) 13.5 cm in front
    (b) .675
    (c) 2.4
15. 4 cm
16. 100 cm, in front of lens, 15 cm high, virtual, erect
17. 24 cm in front of lens
18. 20 cm from lens
19. 1.91 to 1
20. 44.8 cm

Page 249

1. 11.1 cm beyond 2nd lens
2. 2.17 cm
3. 24 in.
4. (a) $f = \infty$
   (b) Same
5. $+12$ cm
6. A
7. 5 cm

8. 11.5 cm beyond 2nd lens, 1.38 cm inverted
9. (a) 17.6 cm
   (b) $+1.67$ diopter
10. 4.17 cm
11. $+2.57$ diopter
12. .323 ft
13. (a) 6.25

(b) 7.25
14. 6 in.
15. 42 in.
16. (a) 30
    (b) 11.14
17. 23.3 cm beyond 2nd lens,

$M = -.444$, real
18. (a) 0.73 ft
    (b) 2.95 in.
19. (a) 50 cm
    (b) 16.7 cm

Page 254

1. .30
2. .314
3. .000059 cm

4. 5.46 cm
5. 5800 lines/cm

Index

Ab-ampere, 184, 187
Ab-henry, 195
Absolute potential, 148
Absolute temperature scale, 122
Ab-volt, 193
Acceleration, 33
  angular, 70
  central, 34
  of gravity, 35, 44
Accommodation, 248
Ac power, 206
Actual mechanical advantage, 65
Alternating current, 204
Alternating electromotive force, 204
Ampere, 160, 164
  international, 164
Ampere-Laplace rule, 186
Amplitude, of simple harmonic
    motion, 82
  of a wave, 107
Analyzer, 254
Angle, critical, 217
  of declination, 176
  of dip, 176
Angular acceleration, 70
Angular displacement, 69
Angular position, 69
Angular velocity, 69
Anions, 163
Anode, 163
Archimedes' principle of buoyancy,
    98
Arm, lever, 26
  moment, 26
Arrow, vector, 13
Astronomical telescope, 245
Atmospheric pressure, 94
Atom, 118
Average velocity, 2
Avogadro's number, 119

Axis, instantaneous, 78
  of rotation, 26

Back electromotive force, 195
Barometric column, 94
Basic concepts, 5
Beats, 108
Beat frequency, 108
Bernoulli's principle, 104
Black body, 134
Body, black, 134
  rigid, 88
Boiling point of water, 119
Boyle's law, 102
Bridge, Wheatstone, 169–170
Brightness, 211
British engineering units, 44
British thermal unit, 125
British yard, 7
Brownian motion, 119
Bulk modulus, 89
Buoyancy, Archimedes' principle
    of, 98

Calorie, 125
Camera, 247
Candle, 211
Capacitance, 152
  electrical, 152
  units of, 153
Capacitive reactance, 205
Capacitor, 152
Capacity, 152
  heat, 125
Cathode, 163
Cations, 163
Center of gravity, 13
Centigrade scale of temperature,
    119
Centimeter, 7

Central acceleration, 34
Centrifugal force, 51
Centripetal force, 51
Cgs units, 44
Change of state, 128
Charge, 140, 153
  electric, 140
  negative, 140
  positive, 140
  unit positive, 141
Chemical effects of current, 163
Circle of reference, 81
Clockwise torque, 27
Closed pipe, 111
Coefficient, of compressibility, 90
  of friction, 17
  of linear expansion, 120
  of mutual inductance, 194
  of rigidity, 89
  of self-inductance, 195
  of thermal conduction, 132
  of volume expansion, 121
Column, barometric, 94
Components of a vector, 14
Compound microscope, 246
Compressibility, coefficient of, 90
Concave mirror, 223
Concepts, 1
  basic, 5
  derived, 5
  of electrical resistance, 162
Concurrent forces, 19
Condenser, 152–153
Conditions of equilibrium, 19, 27
Conduction, thermal, 132
Conservation, of energy, 60
  of momentum, 55
Conservative force, 60
Constant, dielectric, 140
Continuity, law of, 104
Convection, thermal, 132, 134
Conventions, sign, 221
Converging lens, 233
Convex mirror, 222
Coulomb, 164
Coulomb's law, of charge, 140
  of magnetism, 174
Counterclockwise torque, 27
Critical angle, 217
Current(s), alternating, 204
  chemical effect of, 163
  electric, 159
  heating effect of, 161

Current(s) (cont'd.)
  intensity of, 159
  lagging, 205
  leading, 205
  magnetic effect of, 183, 184
Curvilinear motion, uniform, 35

Day, mean solar, 7
Deceleration, 34
Decibel, 113
Declination, angle of, 176
Density, 92
  magnetic flux, 176, 181, 183
  weight, 93
Derived concepts, 5
Diamagnetism, 176
Dielectric constant, 140
Difference in electric potential, 147
Diffraction, 253
Diopter, 239
Dip, angle of, 176
Displacement, 32
  angular, 69
  of simple harmonic motion, 82
Dissipative force, 60
Diverging lens, 233
Doppler's principle, 109
Dynamics, linear, 42
  rotary, 72
Dyne, 44

Effective value of current, 204
Elastic limit, 89
Elasticity, 88
  modulus of, 89
Electric capacitance, 152
Electric charge, 140
Electric current, 159
Electric field of force, 142–143
Electric field intensity, 143
Electric potential, 147–148
Electric resistance, 160
Electrical nature of matter, 142
Electricity, concepts of, 140
Electrochemical equivalent, 164
Electrolysis, Faraday's law of, 164
Electrolyte, 163
Electromagnetic induction, 192
Electromotive force, 160
  alternating, 204
  back, 195
  induced, 192
Electron, 118, 142, 159
Electrostatic unit, 141

Electrostatics, 140
Energy, 59
    conservation of, 59, 60
        heat, 135
    kinetic, 6, 60
    potential, 60
    of rotation, 73
Equilibrium, 19
    conditions of, 19, 27
Equipotential surface, 149
Equivalent, electrochemical, 164
Erg, 148
Expansion, thermal coefficient of
        linear, 120
    thermal coefficient of volume, 121
Experiment, Melde's, 108
External resistance, 160
Eye, 248

Factor, power, 206
Fahrenheit scale of temperature, 119
Fall, free, 35
Farad, 153
Faraday, 164
Faraday's law of electrolysis, 164
Ferromagnetism, 176
Field, electric, 142–143
Field intensity, electric, 143
    magnetic, 175, 181
Figures, significant, 8
First condition of equilibrium, 19
Fluid(s), 88
    in motion, 104
    at rest, 93
Fluid flow, steady, 104
    stream line, 104
Flux density, magnetic, 176, 181,
        183
Flux lines, 176
Focal length, 232
Focal point, 223, 233–235
Foot, 7
Force(s), 44
    addition of, 13
    centrifugal, 51
    centripetal, 51
    concurrent, 19
    conservative, 60
    dissipative, 60
    electric field of, 142–143
    electric line of, 143
    electromotive, 160
        induced, 192

Force(s) (cont'd.)
    of friction, 16
    impulse of, 54
    magnetic line of, 175, 181
    moment of, 26
    nature of, 12
    nonconcurrent, 26
    normal, 17
    parallelogram of, 14
Free fall, 35
Freezing point of water, 119
French Engineering System, 44
Frequency, 81, 108
    of beats, 108
    resonant, 205
Friction
    coefficient of, 17
    force of, 16
    law of, 17
Fundamental, 111
Fusion, heat of, 128

Galilean telescope, 247
Gas, 88, 121, 122
    ideal, 102
Gas law, general, 122
Gas pressure, 119
    Boyle's law, 102
Gauss, 183
Gear, worm, 65
General gas law, 122
Geometric optics, 210, 220
Glass, opera, 247
Gram, 7, 44
Gravity, acceleration of, 35, 44
    center of, 13
    pull of, 13
    specific, 93
Gyration, radius of, 73

Harmonic motion, simple, 81
Harmonic overtones, 109, 112
Heat, 118
    conduction, 132
    convection, 134
    of fusion, 128
    mechanical equivalent of, 136,
        161
    radiation, 132
    specific, 126
    transfer, 132
    of vaporization, 128
Heat capacity, 125

Heating effects of current, 161
Henry, 195
Hooke's law, 89
Horsepower, 60
Hydraulics, 104
Hydrostatic pressure, 90
Hyperopia, 248

Ideal gas, 102
Illuminance, 211
Illumination, 211
Image, real, 223
  virtual, 224
Impedance, 204
Impulse of force, 54
Inch, 7
Inclined plane, 65
Index of refraction, 216
Induced electromotive force, 192
Inductance, mutual, 194
  self-, 194–195
Induction, electromagnetic, 192
  magnetic, 181–183, 185
Inductive reactance, 205
Inertia, moment of, 72
Instantaneous axis, 78
Instantaneous velocity, 33
Instruments, optical, 245–248
Intensity, of current, 159
  of electric field, 143
  of light source, 211
  of magnetic field, 175, 181
Interference of light, 253
Internal reflection, total, 217
Internal resistance, 160
International ampere, 164
International Standard Meter, 7
Ion, 163

Joule, 135
Joule's law of current, 161

Kilogram, 7
Kinematics, rotary, 69
  of translatory motion, 32
Kinetic energy, 6, 60
Kinetic theory of matter, 118

Lagging current, 205
Lambert, 211
Lateral magnification, 221
Law(s), of Boyle, 102
  of continuity, 104
  of Coulomb

Laws, of Coulomb (cont'd.)
  charge, 140
  magnetism, 174
  of friction, 17
  gas, general, 122
  of Hooke, 89
  of Joule, 161
  of Lenz, 194
  of motion, Newton's, 42–43
  of Ohm, 160
  of reflection, 216
  of refraction, 216
  of Snell, 216
Leading current, 205
Left-hand rule, 182, 185
Length, 6
  focal, 232
  wave, 107
Lens, converging, 233
  diverging, 233
  objective, 246
  spectacle, 248
  thin, 231
Lens power, 239
Lenz's law, 194
Lever, 65
Lever arm, 26
Light, 210
  diffraction of, 254
  interference of, 253
  nature of, 210
  polarization of, 254
Limit, elastic, 89
Linear dynamics, 42
Linear expansion, thermal coefficient
  of, 120
Lines of force, electric, 143
  magnetic, 175, 181
Liquid, 88
Longitudinal wave, 107
Loops, 108
Loudness, 113
Lumen, 211
Luminance, 211

Machines, 64
  efficiency of, 65
Magnetic effect of current, 183–184
Magnetic field intensity, 175, 181
Magnetic flux density, 176, 181,
  183
Magnetic induction, 181–183, 185
Magnetic line of force, 175, 181

Magnetic moment, 175
Magnetic permeability, 174, 183
Magnetic poles, north and south, 174
Magnetism, 174
  Coulomb's law of, 174
  terrestrial, 176
Magnification, lateral, 221
Mass, 6, 12, 44
Matter, bulk, 92
  electrical nature of, 142
  kinetic theory of, 118
Maxwell, 183
Mechanical advantage, actual, 65
  theoretical, 65
Mechanical equivalent of heat, 136, 161
Mechanics, 12
Melde's experiment, 108
Meter, 7
Meter, Venturi, 105
Method of components, 14
Method of mixtures, 126
Metric absolute units, 44
Microscope, compound, 246
  simple, 245
Millimeter, 7
Mirror, concave, 223
  convex, 222
  plane, 224
Mixtures, method of, 126
Mks units, 45
Modulus, bulk, 89
  of elasticity, 89
  Young's, 89
Molecule, 118
  random motion of, 118
Moment, of force, 26
  of inertia, 72
  magnetic, 175
Moment arm, 26
Momentum, 6, 54
  conservation, 55
Motion, Brownian, 119
  kinematics of translatory, 32
  Newton's laws of, 42–43
  pendulum, 84
  periodic, 81
  projectile, 52
  rotary, 19, 69
  simple harmonic, 81
  translatory, 19
  uniform, 34

Motion, Brownian (cont'd.)
  uniform curvilinear, 35
  uniformly accelerated linear, 34
  wave, 107
Mutual inductance, 194
Myopia, 248

Negative charge, 140
Neutron, 118
Newton, a force unit, 45
Newton's laws of motion, 42–43
Nodes, 108
Nonconcurrent forces, 26
Normal force, 17
North magnetic pole, 174
Number, Avogadro's, 119

Object, virtual, 242
Objective lens, 246
Ocular, 246
Oersted, 175, 180
Ohm, 160, 163, 164
Ohm's law, 160, 161, 162
Open pipe, 111
Opera glass, 247
Optical instruments, 245–248
Optics, geometric, 210, 220
  physical, 210, 252
Overtones, harmonic, 109, 112

Parallelogram of forces, 14
Paramagnetism, 176
Paraxial ray, 221
Paraxial ray formula, 221
Pascal's principle, 94
Pendulum motion, 84
Period, 81
Periodic motions, 81
Permeability, magnetic, 174, 183
Photometry, 211
Photon, 210
Physical optics, 210, 252
Pipe, closed, 111
  open, 111
Pitch, 113
Plane, inclined, 65
Plane mirror, 224
Point, focal, 223, 233–235
Polarization, 254
Polarizer, 254
Poles, magnetic, 174
Position, 32
  angular, 69

Positive charge, 140
Potential, 147–148, 153
    absolute, 148
    difference in electric, 147
    electric, 147
Potential energy, 59
Pound, 43–44
Pound mass, 7, 43
Pound weight, 43
Poundal, 44
Power, 60
    Ac, 206
    lens, 239
Power factor, 206
Pressure, atmosphere, 94
    gas, 119
        Boyle's law, 102
    hydrostatic, 90
Primary of transformer, 196
Principle, of Bernoulli, 104
    of Doppler, 109
    of Pascal, 94
Projectile motion, 52
Proton, 118
Pulley, 65

Quality of sound, 113
Quantity, scalar, 13
    vector, 13
Quantum theory, 134

Radian, 71
Radiation, thermal, 132–134
Radius of gyration, 73
Random motion of molecules, 118
Reactance, 205
    capacitive, 205
    inductive, 205
Real image, 223
Reference, circle of, 81
Reflection, law of, 216
    regular, 216
    total internal, 217
    of waves, 108
Refraction, law of, 216
    of waves, 108
Refractive index, 216
Regular reflection, 216
Resistance, electrical, 159, 160, 162
    external, 160
    internal, 160
    specific, 162
Resistivity, 162
Resonant frequency, 205

Resultant of vectors, 14
Right-hand rule, 192
Rigid body, 88
Rigidity, coefficient of, 89
Root-mean-square value, 204
Rotary dynamics, 72
Rotary kinematics, 69
Rotary motion, 19, 69
Rotation, axis of, 26
    energy of, 73
Rule(s), Ampere-Laplace, 186
    left-hand, 182, 185
    right-hand, 192

Scalar quantity, 13, 149
Second, 7
Second condition of equilibrium, 27
Secondary of transformer, 196
Self-inductance, 194–195
Side thrust, 184
Sign conventions, 221
Significant figures, 8
Simple harmonic motion, 81
Simple microscope, 245
Single spherical surface, 220, 221
Slug, 44
Snell's law of refraction, 216
Solid, 88
Sound, 111
    characteristics of, 113
    quality of, 113
South magnetic pole, 174
Specific gravity, 93
Specific heat, 126
Specific resistance, 162
Spectacles, 248
Spectroscope, 252
Spectroscopy, 252
Spectrum, 252
Speed, 33
Spy glass, 246
Standing waves, 108, 111
Stat-coulomb, 141, 149
State, change of, 128
Stat-farad, 153
Stationary waves, 108, 111
Stat-volt, 149
Steady flow, 104
Ster-radian, 211
Strain, 88
Streamline flow, 104
Stress, 89
Surface, equipotential, 149

Telescope, astronomical, 245
Galilean, 247
terrestrial, 246
Temperature, 118–119
Temperature scale, absolute, 122
Centigrade, 119
Fahrenheit, 119
Tension, Melde's experiment, 108
Terrestrial magnetism, 176
Terrestrial telescope, 246
Theoretical mechanical advantage, 65
Theory, quantum, 134
Thermal coefficient of linear expansion, 120
Thermal coefficient of volume expansion, 121
Thermal conduction, 132
coefficient of, 132
Thermal convection, 134
Thermal expansion, 120
Thermal radiation, 132–134
Thermal unit, British, 125
Thermodynamics, laws of, 135–136
Thermometry, 119
Thin lens, 231
Thin lens formula, 232
Thrust, side, 184
Time, 6
Torque, 26
clockwise, 27
counterclockwise, 27
Torsional wave, 107
Total energy of random motion of molecules, 118
Total internal reflection, 217
Transfer, heat, 132
Transformer, 195–196
Translatory motion, 19, 34
kinematics of, 32
Transverse wave, 107

Uniform curvilinear motion, 35
Uniform motion, 34
Uniformly accelerated linear motion, 34
Unit(s), 36, 43
British Engineering, 44
British Thermal, 125
of capacitance, 153
cgs, 44
electrostatic, 142
of inductance, 195

Unit(s) (cont'd.)
metric absolute, 44
mks, 45
Unit north pole, 174
Unit positive charge, 141

Valence, 163
Value, root-mean-square, 204
Value of current, effective, 204
Vaporization, heat of, 128
Vector(s), arrow, 13
components, 14
quantity, 13
resultant, 14
Velocity, 33
angular, 69
average, 2, 33
instantaneous, 33
of simple harmonic motion, 82
wave, 108
Venturi meter, 105
Virtual image, 224
Virtual object, 242
Volt, 148, 160, 164
Volume expansion, thermal coefficient of, 121

Watt, 161
Wave(s), amplitude, 107
longitudinal, 107
reflection of, 108
refraction of, 108
standing, 108
stationary, 108
torsional, 107
transverse, 107
Wave length, 107
Wave motion, 107
frequency, 108
velocity of, 108
Weber, 183
Weight, 13
Weight density, 93
Wheatstone Bridge, 169, 170
Wheel and axle, 65
Work, 59
Worm gear, 65

X-component of vector, 14

Yard, 7
Y-component of vector, 14
Young's modulus, 89